On Posthuman War

ON POSTHUMAN WAR

COMPUTATION AND MILITARY VIOLENCE

MIKE HILL

University of Minnesota Press

Minneapolis

London

Published by the University of Minnesota Press
111 Third Avenue South, Suite 290
Minneapolis, MN 55401-2520
http://www.upress.umn.edu

ISBN 978-0-8166-6089-6 (hc)
ISBN 978-0-8166-6090-2 (pb)

A Cataloging-in-Publication record for this book is available from the Library of Congress.

Printed in the United States of America on acid-free paper

The University of Minnesota is an equal-opportunity educator and employer.

UMP BmB 2022

For James E. Ullom,
1950–2020

Let Venus have her way,
and she will bring you Mars.
— Henri Bergson

Contents

Acknowledgments

My first debt of gratitude goes to my students at the University at Albany, SUNY. For some years they've been hearing about war in its many forms and as related to matters far beyond the already scaled-up focus of this book.

I have also had the good fortune to share fragments and thoughts-in-process at university talks and seminars from China and Australia, to Finland and Portugal, and back to any number of academic society conferences in the United States. Wang Fengzhen, Min Zhou, Gareth Griffiths, Michael Griffiths, and Benita Heiskanen have all been gracious and supportive hosts. Thanks to listeners and questioners long past. More recently, let me give a quick nod to members of the Re: Enlightenment Project. This group has provided a remarkable laboratory of virtual colleagues and helped me chart future work.

Jeanette Altarriba, dean of the College of Arts and Sciences at UAlbany, granted me a one-semester paid sabbatical to complete the book in 2019. As English department chair, Charles Shepherdson helped usher the process through. Several colleagues in the Department of English gave advice and encouragement along the way: Richard Barney, Helen Elam, and Tom Cohen.

Beyond UAlbany, thanks go to Devoney Looser, Clifford Siskin, Modhumita Roy, Kevin Frye, my one-time and maybe-someday-again collaborator Warren Montag, Jeffery Williams, Paul Lauter, Paul Smith, Henry Sussman, and Cary Wolfe.

At the University of Minnesota Press, I thank former editor Richard Morrison and, most of all, Doug Armato. Doug inherited the book and stuck with it, even though I fell further behind than either of us would have liked. I appreciate the editorial help of Zenyse Miller at the Press. Independently, Stephen Leon read an initial draft of the manuscript and spent hours helping boil down the prose. I also benefited greatly from revision suggestions offered by Greg Lambert and other anonymous reviewers.

USMC Sergeant James T. Marchetti was my "recruiter" for the Parris Island experiment. I appreciate the access more than I was able to express in this book.

Thanks, Mom, for being aboard this long, strange trip called being an English professor. It's still alien to us both.

Most of all, thank you, Katharine Raley. Not only did you read the manuscript all the way through, offering invaluable corrections and insights on a rough and not-quite-ready text, you provided the kindness and care necessary for me to get on with the task.

Finally, my close friend and comrade James Ullom died of Covid at the very end of my writing this book. He was an incredibly smart, humble, and funny old guy, rare combinations in my neck of the professional woods. Thanks, Giacamo. This one is dedicated to you.

Abbreviations

AAA	American Anthropological Association
CE-COIN	cyber-enabled counterinsurgency theory
COIN	counterinsurgency theory
DARPA	Defense Advanced Research Projects Agency
DI	drill instructor
DOD	Department of Defense
GWOT	global war on terror
HTS	Human Terrain System program
MENPs	magnetoelectric nanoparticles
OMB	Office of Management and Budget
OSS	Office of Strategic Services
RMA	Revolution in Military Affairs
SOMA	Stochastic Opponent Modeling Agent
SOS	system of systems
USMC	United States Marine Corps

Preface

Supping with the Devil Dogs

We Make Marines.
—USMC boot camp signage

U.S. Marine Corps Recruit Depot
Parris Island, South Carolina, USA
0400

Arriving at boot camp, nobody said a word. We were squeezed in tight, two-by-two, in a double line of row seats in an old school bus converted to transport young men and women to their first day of becoming U.S. Marines: the receiving phase. We were not recruits, but a group of educators and journalists embedded to experience an abridged form of the training. As a supplement to traditional forms of military recruitment, the Corps also has an interest in "connecting educators with the future of the USMC." This immersive and intensive program accepts physically capable civilians who work with traditional college-age students potentially interested in a military career. I was not there with the intention of recruiting my students. Rather, by experiencing boot camp in short form, my goal was to learn something applicable to the book before you now. It was still dark as we crossed the Beaufort River; I was keen our first morning to get as close as possible to the already recruited. As we passed under the archway and onto Parris Island, first light caught the slogan in Marine Corps yellow and blaze-orange, as if to welcome us to the military side of an educational looking-glass: We Make Marines.

We would not be made into Marines. But our receiving phase followed the same scream-in-your-face direct orders, run-until-you-drop commands, and confusing questions hurled at maximum volume received by real recruits. The Drill Instructor (DI) screamed eye to eye at the distance of a hat brim, "Did I give you permission to brush away my sand flea?" A sheepish older member of our group replied, "Sir, No Sir!" The DI responded, "Scream louder!" "Sir, No Sir!" And on it would go, until the dreaded last straw: "Now get in my sand pit!" Everything on the island belonged to the DI, even things not normally belonging to anybody: the bus we rode in on; the sand pit where people vomited after one too many burpees; the sand fleas we let bite us under the threat of still more burpees; the iconic yellow footprints painted with exacting symmetry for us to muster on at dawn. By the end of the receiving phase, no one could ignore the message. Our "reception" was double in nature: The DIs received us, and in return we got a bitter taste, some literally, of the transformation promised by the signage: We Make Marines.

If we were not going to be made into Marines, as we knew, and as did the DIs, then what was happening in this bewildering scene? We were no longer the writers and teachers we left back home, though the identity spit into our faces just before each command was: "Educator . . . !" Coming from the DIs, the title no longer afforded the life of the mind. Too many years as a humanities professor should have kept me away from anything like this. Could the activity of cultural analysis, discussions about identity and kinship, theorizing about knowledge, and long days in the library mingle with combat training? Asking myself these kinds of questions, the ivory-towered tendencies of my educator-self soon disappeared, and the learning became secondary to the doing.

Parris Island was claimed by a French Huguenot expedition led by Jean Ribault in 1562, the first Europeans to colonize the territory. The island's second invasion force, an English military expedition led by Sir Francis Drake, raided and burned the French colonies nearby. After coming under English control as Port Royal Island, it was granted to the plantation owner Lord Proprietor Robert Daniell as part of the British slave economy, growing indigo and, later, the most lucrative cash crop, cotton. During the American Civil War, Parris Island became a refuge for freed slaves. Thanks to freedman-turned-representative Robert Smalls, the island hosted abolitionist schools. In addition, he established the first military installation there, the original mix between educators and war. Now, in 2021, about seventeen thousand Marine recruits are trained on Parris Island each year. This much I learned from books. But book learning wasn't useful during my time at boot camp. From the receiving phase on, when hailed by the DIs, we educators simply snapped to attention, gave our "battle rattle," and proceeded with the agreement: train on Parris Island approximating the trials of a normal recruit, then go back home and read or write whatever you want.

The DIs were clear about what happens in the process of "making Marines": Whatever civilian identity you had before you got on the island, forget it. The civilian is what's unmade, and it was not as hard as you'd think. This is dramatized not only by the famous haircut scenes depicted in the movies (Kubrick's *Full Metal Jacket* features the same obstacle courses we were hounded into ascending), or by the brute physicality of it all, but there is also a *de*-civilianizing aspect to Marine Corps language.

A favorite command from our DI was where (and when) to look: "Look at me right now!" we'd hear over and over. *Keep me in your sights and you'll know automatically when and where to target something else.* This was how I interpreted her obsession with where we looked. And it was always *right now.* In the same way everything belonged to the DI, for the recruit there is no past or future on the island. Or at least, there is no time at which he or she will either be a Marine or someone having failed to be made into one. Marine Corps time is the eternal *right now,* befitting the Forever War, and the never-ending wake of 9/11. A related favorite slogan, a veteran told me after, is: Once a Marine, Always a Marine. But before "once," and in the midst of our new sense of "always," the past is reverse engineered. For a split second on the way home, at the airport during preboard, long lines, sluggish confusion, passengers roaming in a crowded daze, I thought: *Why don't we stop whining and get into formation?*

A former educator of mine from graduate school, Michael Sprinker, warning against spending time with neofascists for my previous book, *After Whiteness: Unmaking an American Majority,* used to say: "If you sup' with the Devil, you're gonna to need a long spoon." This is not to make a flat equation between these two groups. But there was a similar feeling of losing academic distance from the research at hand. I realize now what I would say to Michael about the Marines, if he were still alive: "Their spoon belongs to the DI. And it's not the Devil, but what Marine's call themselves: the Devil Dogs."

Our remembering to be educators at least gave us a way to not "sup" with the Devil Dogs, in the event we got seriously hurt, or surmised physical harm to be imminent. But in the same sense of the eternal emergency befitting a new kind of war, permanent and proximate, we were always about to get hurt. Being eggheads and supping with the Devil Dogs gave us an identity approximating what the DIs wanted to "make" in a new brand of Marine, what they call Deviled Eggheads.

For the devilish part of our training, there was the high-wire obstacle course, the ungodly predawn hours with too little sleep, the sand pit and the DI's uninhibited sand fleas, the relentless drilling, the gas chamber (twenty-odd educators with grimaced leaky faces), and the rifle range (I was embarrassed in front of the eggheads to shoot well at 200 yards), and finally, dropping through the "hell hole" down the one hundred–foot rappel tower. During our training, success depended on the "team" and not the "self." This self-effacing message was repeated by our DI during unusual bits of downtime, gathered on the barracks floor, in line for "chow," or on the way to a "head check" (the bathroom). In those moments nobody thought selflessness was trite.

The release documents were signed and notarized, along with medical papers documenting our physical condition. This reassured the Marine Corps that we could be pushed a certain distance, just the right amount of emergency, without legal recourse against the U.S. government in the case of major injury or death. Any educator could "tap out." But the release forms signified a different kind of "release" than merely a physical one. Crossing over from civilian life to fantasy boot camp meant you didn't have to "tap out" in an emergency. Unlike in the civilian world, team members put aside their pain and rallied each other through.

There is no mind/body split on Parris Island. But the unmaking of the civilian in the purely corporeal sense was not the most significant part of how to "make Marines." In the manner of signed release forms as much as by governmental acts, identities are made and unmade by the law. Civilian-suspects detained under the U.S. Patriot Act, or held in prisons overseas, know this lesson better than most. Just after mustering on the Yellow Footprints, and following a wicked dose of sprints, we were hustled by a swarm of more DIs, chest to the back of the educator in front of you, into a briefing room. We were not allowed to use the stainless-steel doors over which a sign said only the "finest fighting force in the world" could pass through. We passed through side doors, reserved for civilians, even though we marched in double time like no civilian ever would. Once inside, under the florescent lights and on the shiny red waxed floors, we learned about a different kind of decivilianization. It was another instance of "release," this time having to do with the duties of the soldier and a loss of civil rights.

In his thunderous way, one of the DIs informed us we were no longer subject to the laws of the U.S. Constitution. Instead, we would be held accountable by the Uniform Code of Military Justice. As I realized later, our shuffling in double time led to another kind of duplicity, once more highlighting the intermingling between citizen and would-be combatant. Being made into a Marine will give you the virtues of collective thinking and self-sacrifice, but with one qualification: while you have those virtues, you are no longer a civilian. We were still really just educators, and I don't know whether our civil rights were actually suspended, or if reference to the Uniform Code of Military Justice was to further dramatize our participation-observation. But within the eternal Marine Corps "now," what was drama and what was real hardly mattered. Refusal to obey a direct order is a legally punishable offense for the real recruit.

This got me thinking about a different kind of obedience, the social kind, which doesn't depend on a direct order but an indirect one: the adherence to civic norms and the law. There is a famous dictum along these lines declared by Immanuel Kant (1724–1804): "Argue as much as you like and about whatever you like, but obey!" Kant was speaking on behalf of the use of public reason, freedom of the press, religious independence, and the usual list of rights guaranteed to the members of a representative democracy. But his command, "Obey!," with an exclamation point coming right out of boot camp, is equally important. *Talk all you want. But you cannot disobey sovereign law and legally maintain your status as a member of the state. As a citizen, the law is an extension of you: To betray it is to become one of our foes.*

We have two different forms of obedience, then: the warrior's adherence to command, and the civilian's adherence to the social and legal rules required to be a member of the *polis*. The second form of obedience is necessary for what we call freedom, Kant says; and as the patriotic slogan goes, the first form is essential for protecting the "obedient" free. The civilian is on one side of Kant's great divide, the side of individualism, peace, knowledge, and voluntary association. While on the other side is the "team," the mission, direct orders, and the chain of command. Such an arrangement, the bedrock of political modernity, would seem clear enough, even with the paradoxical idea of "obedient" freedom.

What I experienced as an educator turned faux Marine recruit broke the rules of Kant's divide. But I am not the only one breaking the rules. This book explores the expansion of war along the lines of decivilianization in a larger sense, scaled up from my "supping with the Devil Dogs" to the unseen ways military violence enters ordinary life: the loss of civilian identity; the application of new media for purposes of war; the weaponization of thought. I'm interested in how civil society activities like these comingle with combat operations, and how the process of decivilianization has expanded in the form of "posthuman war." By this I mean to draw associations between a new kind of military violence and those questions we tend to think about as emerging from within the peaceful realms of civil society: Who am I? With whom do I belong? How do I know?

Each of the three chapters to follow thus takes an individual discipline—demography, anthropology, and neuroscience—and shows how they are applied in posthuman war to these very human questions. Identity, social organization, and cognition are the issues I address as they have crossed Kant's great divide. When I think of scaling up my own experience of decivilianization in training with the U.S. Marines, the enigma of "scale" is itself worth pausing over. Even if the end of civil society described in my analysis has been missed by most civilians, the invisibility of war doesn't mean it's not ubiquitous. "You may not be interested in war," Leon Trotsky (1879–1940) said, "but war is interested in you."

We often miss large-scale phenomena, since our social and intellectual tendencies portray reality as reducible to what we think we already know. This is why I draw on computation and realism. Scaling up requires new technologies, useful in war and peace, and is essential for explaining reality better. Traditional forms of opposition are blurred by more massive relationships, which are occulted by their complexity, but are also actually

there. We are witnesses to the disappearance of standard political and social divisions, not just between war and peace, friend and foe, citizen and
soldier, but also between what exists and what's rendered visible by new
media.

Moving through the three fields of study identified in what follows
as war disciplines (demography, anthropology, and neuroscience), computational technology is key. Computation is key because, like war, it is
both ubiquitous and largely invisible. But in focusing on largely invisible
wars, I don't want to leave the reader thinking: *Why all this theory about
unseen wars, when the predominance of military violence is obvious, as any good
account of history will show.* Indeed, war is ostensibly a permanent part of
United States history. Since 1776 (and before, if we include Puritan acts of
aggression), the United States has been at war for 227 out of 244 years; 93
percent of the country's existence. So why a book about the wars citizens
don't see?

It's a fair question. Where relevant, I discuss more obvious acts of
military violence, like the removal of Native Americans, the internment of
Japanese citizens, the Patriot Act, and other instances where Kant's great
friend–foe divide becomes blurred. But the point of this book is to link the
visible wars of bullets and bombs with the invisible ones of virtuality and
data. War is not an anomaly of the Kantian state but the secret condition
of the state's existence. War's public invisibility prepares the ground for its
political affirmation.

Recall President Trump's declaration during the "American Spring" of
2020: "We're doing it in Washington, D.C. We're going to do something
that people haven't seen before." The "it" in question was to "dispatch
thousands and thousands of heavily armed soldiers, [and other] military
personnel to U.S. cities" in response to public demonstrations triggered by
the death of George Floyd. Floyd, an unarmed black man killed by a white
police officer who knelt on his neck during an arrest for the alleged use
of counterfeit currency, became a rallying point against the killing of innocent black citizens, specifically by police. This death, resulting in the officer's conviction of three crimes, including "second degree unintentional
murder," inspired the largest antiracist demonstrations in U.S. cities since
the 1960s. And yet the appeal to civil rights in the current epoch solicited
military activation *within* the state, and not simply in opposition *against*

other states. The war made visible by the demonstrations was not the one I saw by way of armed troops in my hometown. It was the longer war of racist violence.

Trump criticized a small number of opportunistic criminal acts associated with the protests, using the inflammatory term "domestic terror." His attorney general likewise asserted an affiliation, nowhere to be found, between the Black Lives Matter movement at the head of the unrest and "foreign actors playing all sides to exacerbate violence."

The assertion of a hostile force hiding within the state is consistent with the Global War on Terror, and is another feature of posthuman war: the battlefield extends from all points abroad into the so-called U.S. homeland. Today, U.S. citizens live within USNORTHCOM, a new zone of expanded military command. Borders demarcate national boundaries, with Canada to the north and Mexico to the south. But USNORTHCOM's borders are the internal boarders seen only by the war planner's mapping machines. Ten carefully circumscribed intranational sectors divide the country for future military operations. These operations can occur, paradoxically, both for and against the nation's citizens. Here a "new kind of enemy" mixes in and out of civilian life. In the context of the coming U.S. white minority, one strategic plan for suppressing civil disturbances names "large numbers of minority groups" as the most likely flashpoint of violence. USNORTHCOM does not require a formal declaration of war to enact preemptive strikes on the civilian side of Kant's citizen–combatant divide.

As I write this preface in mid-2021, the Federal Bureau of Investigation has declared racial and ethnic tensions to be the most likely source of extremist violence in the near future. As the January 6 insurgency in Washington, D.C., might suggest, the oscillation between civilian and a new kind of soldier, like the oscillation between educator and faux Marine, is easier (and perhaps also less new) than we might like to think. What follows is an invitation to think of war both differently and once again.

Introduction

Number Rules

> *Rule Number One: Terrorists are human beings. They are human beings with emotions that can be channeled into lethal action and often bring innocent people within their definition of "enemy." If you ignore them as people, you may witness the horrifying determination of human intelligence.*
>
> —*Terrorist Recognition Handbook*

THE *TERRORIST RECOGNITION HANDBOOK*

As stated in this epigram, rule number one of terrorist recognition is simply: Terrorists are human beings.[1] But might it be useful to adjust the logic of this statement, and ask if such a rule works in the reverse? How does terrorist recognition turn humanity into a category so porous as to be fatally inclusive? If we were to think about humanity as encompassing terror, rather than just the terrorist as human being, then we would have to address how terrorist recognition involves the concept of "number" at its very core. Here the dichotomy of one identity existing in lethal opposition to another gives way to a more nettlesome problem, a problem not of *the* enemy, but of *enemies,* in a horrifyingly capacious sense. The growth of the enemy's numbers is the more occulted rule the *Handbook* says we ignore at our peril. Thus rule number one intimates a second rule, with the message of both rules being the need to get beyond the very concept of the two: not simply us-*versus*-them, but a second, more precarious reality, and one perhaps more to the point of the enemy's ubiquitous growth. War is no longer about us-*versus*-them, but us-*as*-them, instead.

The 2018 *National Security Strategy (NSS)* describes "an increasingly complex global security environment" where "humanity" is redefined in similar ways, permeated by "terrorism," indistinguishable from "enemies," and subject to new forms of military recombination. The *NSS* might be described as a founding document of posthuman war because of its emphasis on the enemy's numbers.[2] For example, it stresses the importance of "rapid advancements in computing, 'big data' analytics, artificial intelligence, autonomy, robotics, directed energy, hypersonics, and biotechnology" (3). Computation, broadly conceived as the knowledge produced by individuating, counting, and reassembling what we know about the world and how we relate to one other, has reached a level of development where human beings are not simply to be targeted by or protected from war but are instead reconceptualized as entities infused with military violence. The statement "Technologies will ensure we will be able to fight and win the wars of the future" parallels a second statement: "Our Homeland is no longer a sanctuary" (3). "Futurity is no longer" determined by having a time or a place without war, even when such a time and place involves personal, interpersonal, and cognitive activities. What we might call the *de-civilianization* of civil society is how U.S. security doctrine makes civilians secure. Paradoxically, within this form of security there is no end to war, no place too close to home, or too far from the battlefield, to escape the reach of military violence.

To say that war extends into the formerly neutral territory of the human being per se is not to disregard the historical record. In presenting a book on the anthropological (chapter 1), demographic (chapter 2), and neuroscientific (chapter 3) applications of war, I do not wish to diminish the myriad examples of how war invading the lives of noncombatants in the more distant past (Native American genocide, conscription, Japanese internment, violence against citizens by state police and federal troops, the millions upon millions of civilians who have been killed or wounded by military actions).[3] My focus, however, is on a recently activated and specific kind of war: war as manifest within the human domain, our individual, sociocultural, and biological existence, and how computational intelligence takes part in the expansion of military violence within these zones during the first decades of the twenty-first century.

The U.S. Department of Defense's change of emphasis following 9/11, from "blunt force" to the "smarter operations of Command, Control, and Communications, or C3" (6), is further evidence of the "terrorists are human beings" reversal. Since information exchange is the ultimate civil-society activity, communication takes on special importance in what I call posthuman war. The three Cs listed by the *NSS* call forth a link between war within the "human being" and the breakthrough war discipline of cybernetics, originating in World War II. In both instances, "communication" extends to all information systems. Given the overlap between humanity and technology already activated by a C3 approach to war, the *NSS*'s appeal to smart war is intelligence-*as*-violence in the deeply subjective, as well as computational, sense. To offer such a hypothesis is not to suggest human intelligence has ever been nontechnological. To the contrary, in the practice of posthuman war, all intelligence is presumed to operate within one media form or another. Moreover, to push the C3 approach to war to its current extreme, all intelligence operates as media.

As each of the three subsequent chapters presented in this book describe, the battlefield-as-data-field now includes personal, cultural, and cognitive interaction in this computationally oriented way. "Bullets, Beans, and Data" make up the logistical focus of the New Army, with an emphasis on how data reorients what counts as war matériel.[4] "Show Me the Data," one subheading from the "All Domain Operations" doctrine reads ("BBD"). Here, as in the *Terrorist Recognition Handbook*, number rules. To wage war in the twenty-first century means not just being an expert in weapons traditionally defined but also to take part in the technical redefinition of what it means to be a human being understanding the world. The epoch of posthuman war ushers in the rise of the "warrior intellectual," so named to emphasize academic training in the human sciences as foundational to General David Petraeus's all-domain approach to counterinsurgency theory.[5] War in this context must be viewed as an episto-military art.

As I discuss in chapter 1, the much-hyped Revolution in Military Affairs (RMA) inspired after 9/11 has redefined individual human identity as an operationalizable part of the battlefield. In chapter 2, I show how the RMA has extended from an identity-based emphasis on the individual identity of insurgents to broader-conceived relations of community, or "culture." Chapter 3 offers an even more intimate focus, following posthuman war's movement inward from these subjective and intersubjective domains toward the similarly operationalizable cognitive realm of the human brain. All three areas of focus—the individual, the communal, and the biological—are addressed in RMA thinking as enmeshed within an overlapping array of informational frontiers. Humanity not only utilizes "communication, command, and control" in the way the cyberneticist thinks about information, but humanity is also reconceptualized as a complex computational entity in its own right. But this is not a book, like so many others, about war as the rise of machines. To the RMA way of thinking, humanity has always been machinic.

A *U.S. Army Special Operations Manual on Unconventional Warfare* document puts the cybernetic mind–media–matter connection clearly when insisting: "Every soldier is a sensor first."[6] Every soldier is a sensor, and in the same way number rules in the *Terrorist Recognition Handbook*, the sensing and the soldiering are of a piece in posthuman war. Along with the opposition between friend and foe, the distinctions between culture and biology, human and nonhuman, animal and machine, and even reality and representation, are subject to overlap, reversal, and recombination. If war matériel includes bullets and beans, as well as data, then it is because the New Army runs on computation: media equals matter—equals war. Such a statement has profound implications for the way in which foundational war categories are being highlighted, demolished, and reorganized at more expansive levels. As the *NSS* notes further, "Rogue powers exploit ambiguity, and deliberately blur the lines between civil and military goals" (2).

The ambiguity activated by identifying the enemy-as-rogue ushers in a form of violence existing within what were formerly peaceful relations. These are relations not just between, or even within, states, but within social as well as intimate behaviors. The curious notion of the "soldier as sensor" is partly explainable by recalling Michel Foucault's theory of biopolitics. Here the personal and political elements of human life are securitized by the state along expressly technical (indeed numerical) lines. In the shortest formulation possible, biopolitics is Foucault's generally well-known term for describing "the medico-normalizing techniques" connecting an individual's biological, social, and geographical markers to modern forms of sovereignty.[7] Biopolitics is therefore also connected to Foucault's theory of "governmentality," in Colin Gordon's useful synopsis, a form of "modern governmental rationality, [which] is simultaneously about individualizing and totalizing . . . [,] about finding answers to the questions of what it is for an individual, and for a society or populations of individuals, to be governed or made governable."[8] For Foucault, human society is originally conceived as in need of securitization.

In posthuman war, identity, population, and medical discourse are intertwined along Foucauldian lines, which is also the line of organization followed by this book as the three chapters move from demographics to anthropology to neuroscience. But there is a difference between my account of posthuman war and Foucault's oft-cited remark: "Society must be defended." The kind of war I have in mind is not merely social; nor is it merely what states do against each other as "politics by other means"; nor is it limited to one side of the battlefield as determined by the reason of state. Rather, I am interested in "means" themselves as multipliers of military violence. A key component of what some military theorists call World War X is the weaponization of X-factors formerly thought to be disassociated with military violence.[9] The X-factors of posthuman war include universal attributes—or at least very large-scale ones—including the definition of the humanity, the primacy of communication, and the way reality is understood. To feel one may differentiate oneself from another in any pure sense of such a division belongs to a different time and place. The posthuman military field manual will tell you war is no longer a matter of black and white, but is instead so many different shades of grey. The better technical term, because it applies to both battlefields and brains, is "grey goo." Thus, *NSS* also tells us, the "post-WWII international order" means opening "new fronts" within the larger context of a "violation of the principles of sovereignty" (2).

Posthuman war is *autogenic*, meaning two things: military violence is not simply an application of state power in overtly visible ways; and its victories function to promote war's endless renewal. Such a "violation of sovereignty" initiates a change, equally "terrifying," in what Roberto Esposito calls the "protection and negation of life."[10] In David Kilcullen's chilling conception, humanity at large is turned into "accidental guerrillas."[11]

Widely available in its second edition, the *Terrorist Recognition Handbook* addresses "you, the professional" (7). But this kind of professionalism is open to both being and knowing the rogue. The *Handbook's* readership is composed not only of civilians-as-enemies within but also various kinds of "police." "Customs officials, security managers" and other law-enforcement professionals are part of the book's audience, anybody whose work demands the latest in "terrorist education" (3). Yet this is according to conditions where enemies are indefatigably proximate and pluralized. If we "consider everyone a potential terrorist," then "every person . . . even the casual observer . . . is also a potential surveillant" (189)—and vice versa. We the people are both threatened by political violence and regarded as its orgin. Some future terror of this schizophrenic sort loiters imperceptibly beneath the surface of "humanity's" disguise (225). This is "the nature," the *Handbook* suggests, of "using ethnic or territorial rhetoric as a 'political' mask" (24). But behind the political mask are only more disguises, each one perpetuating the computational spiral of a never-ending security breach.

Judith Butler's work on "precarious living" is useful in naming this challenge to civilian existence: how the ontological, social, and biological boundaries once delimiting our conceptions of humanity are both erased and reconfigured as commensurate with military goals. Butler evokes the word "precarity," meaning "the political implications of those normative conceptions of the human that produce, through an exclusionary process, a host of unlivable lives whose legal and political status is suspended."[12] The reference to "normativity" is consistent, to a point, with posthuman war's expansion into identity, culture, and cognition. Along the lines of Georgio Agamben's influential notion of a permanent state of emergency, my focus on these three areas should add to previous knowledge about nonnormative ways of being as precisely the new norm.[13] But if number rules in posthuman war, then we must reconceptualize the notion of difference-as-opposition and replace it with an emphasis on quantitative complexity more akin with war's new means. We must take an additional step from Butler's still-too-narrow focus on precarity as a principle of exclusion. To understand posthuman war, we must emphasize the *inclusive* processes by which humanity is counted, categorized, and reassembled as part of the military arsenal. The question remaining to be answered is not just who is left out, but also, who (and what) is moved in, when war disregards the demarcation between homeland and battlefront, and redraws the boundaries, apropos cybernetics, between mind, matter, and machine. Living in a context where "terrorists are human beings," where the enemy is everywhere-and-nowhere all at once, and where the identification of the rogue means the rogue is also me, portends a state of war inclusive of all that war used to exclude.

At its furthest reach, we could say that the "enemy's" proximity-in-numbers upsets modernity's whole oppositional applecart, from Western philosophies of being, to traditional ways of recognizing each other, to our understanding of what constitutes reality. As with rule number one of posthuman war, the friend-*versus*-foe distinction disappears when the battlefront and home front merge together. But note, too, the further extension of rule number one: alongside its fading political hopes, humanity's ontological and epistemological status also disappears. Brian Massumi's term "ontopower" helps unpack posthuman war's triple (ontological, social, and epistemic) threat as military violence invades traditional conceptions of the human race. Massumi describes "an operative logic of power" working in an "infinitely space-filling and insidiously infiltrating" way.[14] On the one hand, ontopower jettisons older forms of boundary-making (between subject and object, animal and machine, media and matter), holding humanistic ideology intact. On the other hand, the "post" in posthuman should suggest not a simple move backward in human evolution. It should suggest instead a species-altering move toward a future full of maximal freedom and risk. Because posthuman war creates hybrids comprising data-substance (bullets and beans equal data), the autogenic nature of posthuman war performs resistance *to* resistance. How else can we be both citizens and rogues simultaneously, the lesson of rule number one? How else does the protection of humanity become synonymous with the reality of its annihilation, to follow the extension of this rule?

Bernard Stiegler's *Technics and Time* is useful for surmising posthuman war's inclusive nature, and can be read alongside Foucault's notion of "technologies of self." Stiegler connects the five-hundred-year-old devaluation of "technicity" to Western philosophy's original sin. "At the beginning of its history," he writes, "philosophy separates *tekhne* from *episteme*."[15] This division enables a second, still more stultifying dichotomy, opposing "beings" and "things" (1). The separation between humanity and matter,

Stiegler suggests, is connected to "the Industrial Revolution," as well as "the disciplines of ethnology, sociology, anthropology, general history, and psychology" (2). It produces what Stiegler calls an "eidetic blinding" to the fuller and more accurate vision he wishes to reveal: the persistence of "technical being" (2). Posthuman war is made possible by the rebirth of "technical being." This is because the latest war technologies no longer have the "eidetic blinding" they once did. Military violence no longer separates "being" from "thing," and has advanced instead in accordance with C3's alternative episteme: "all domain warfare," a new reliance on data, which allows the joining of subjects and objects in more open-ended and inter-penetrating ways. The close connection between ontology and knowledge, and the subsequent one between knowledge and the material world, is ultimately at stake in Stiegler's affirmation of *tekhne*. The triple project of Western modernity, its way of identifying individuals, categorizing social groups, and understanding reality, is reoriented by posthuman war through computational means: *tekhne* without "eidetic blindness."

A Department of Defense (DOD) document called *Network Centric Data Strategies* unveils a range of Stiegler-like fusion tools designed to "pull multiple sets of data together so as to create a current picture" of this larger human battlefield.[16] As indicated by the word "flexibility," when netcentric war practitioners talk about a "many-to-many" paradigm for receiving and transmitting data, it is obeying the same number rules as the *Terrorist Recognition Handbook* (2). It is also an example of abandoning the being *versus* thing relation Stiegler sees as the crumbling foundation of Western modernity. The word "current" in the DOD document indicates how netcentric technologies initiate a new space–time arrangement, working in a virtual way: What is invisible is exactly what is really there, much like the "terrorist rogue" in "humanity's" disguise; and what is really there exists at a scale of data incomprehensible for traditional forms of human intelligence. In the sense that DOD embraces a C3 model as consistent with the concept of technical being, we might call posthuman rouging: Cybernetics 2.0. Relations, not identities, are what fusion tools render knowable, whether they become weaponized or not. According to the DOD's data goals, the word "mediation" takes on special meaning. Media here denotes a way not merely of representing targets but also of creating them through the representational act: "Mediation resolves the differences in name, structure, and representation of data" (9). Computational resolution of this sort both "discovers data assets" and "makes data visible" (9). What is there and awaiting discovery must also be produced. But more mediation equals more layers of disguise—more data put together in ways that can expand the possibilities for military violence.

CYBERNETICS AND NETCENTRIC WAR

The famous Clausewitzian paradigm of war as "a duel by other means" is all but gone in posthuman war. This is because war's means are no longer distinguishable from its ends. Both are eternal, and if we think of means as meaning media, then the perpetuation of war must also signal the displacement of traditional categories of information exchange. With the displacement of war's dualisms by network-centric coordinates, gone are the ideals of communicative reason transcending mind, matter, and machine.[17] We are practicing war today by other-than-usual means because the means are coequal with a universe of data patterns where human relations become war matériel. The means of posthuman war function by twisting Clausewitz's duels into more fluid, complex, and multifarious relationships. These relationships absorb previous war dualisms, not only oppositions on the battlefield but also the many binaries underwriting the human being per se (mind/body, human/nonhuman, means/ends, matter/media, war/peace). Identity, social organization, and cognition, posited by Western idealism as cordoned-off from war, are now the way war is made.

Netcentric war thus seeks to exploit the following: (a) computational technologies for rendering a new array of targets visible when they cannot be identified by human intelligence alone, (b) the fragility of civil society practices insofar as they no longer exclude war, and (c) the hybrid identity of war fighters moving without friction between peaceful and violent domains. In thinking further about netcentric war, the term "mediation" is worth further explanation. The term helps further introduce how my three war disciplines—demography, anthropology, and neuroscience—are featured as examples of posthuman war. "Mediation" covers minds and matter, bodies and terrain, as well as people and the law. The strategists of "irregular" war thus draw upon innovative data applications to render visible myriad overlapping fronts within the largest possible frame.[18] The key point from cybernetics is that the applications themselves share an informational dynamic with the entities they connect with. The *regularization* of "irregular war" is achieved through the connection to connections, overlapping data systems, where data and system refers to human, material, and media technologies alike. The machines used for discovering a target, the targeted, and soldier doing the targeting, cross over into one another as a relation of "many-to-many." Here oppositional relationships are again replaced by a network-centric information-based dynamic. Extending the rule of numbers all the way through: There is no distinction between "data definition and its structures," no difference between "community ontology and data organization."[19]

Under the all-accommodating heading of "data ontology," posthuman war gains its ultimate ground. The relationship between human beings, things, and knowledge is redrawn along what cyberneticists and military strategists recognize as the lines of "command, communication, and control." We might be reminded of the full title of Norbert Wiener's watershed book, *Cybernetics, or Control and Communication in the Animal and Machine.*[20] In Wiener's subtitle, the word "or" designates a unique moment in the history of science where human and machinic forms of intelligence not only share operational dynamics but also overlap. "Communication" is presented here as a "control" problem, and more important, as a "fusion tool" organizing "big data." But the size of the data being controlled is more significant than often recognized. Information scaled-up beyond its usual confines reveals a common if occulted bond between "animals and machines." If we add numbers to existing categories, we can expect those categories to change. Categorical change through computation is the more general lesson of saying number rules in posthuman war.

The advancement of computational epistemology, which includes but goes beyond the literal sense of computers, mediates war's expansion into humanity's subjective, communal, and epistemic zones. But in saying so it is important to avoid the tempting syllogism: war uses technology; war is bad; therefore, technology is bad. Nothing is more human than the use of artificial tools, and book writing itself is clearly a form of media technology. There is plenty of technology critique, going back to an old division between technology and aesthetics, that claims computation "ignores the social and material character" of human life.[21] It is no doubt important to think about the relationship between "organized networks and non-representative democracy."[22] But it is impossible to think about democracy existing with no technology at all. A recent anthology on organizational science lists Descartes and Galileo, Locke and Newton, Bacon and Kant, Spinoza, Leibniz, and Hume as joined together in starting Western Europe down the path of "proposing and testing quantitative theories and models of natural law."[23]

The objection made to "quantitative" reason in this volume claims that computers equal "control technology," where "control" tends exclusively toward inhumanity, repression, and death. These critics declare, "The eighteenth-century's . . . new emphasis on [technical] precision in measurement" marks a "quantificational spirit" constrictively "normative" at its core. But anybody interested in the promises of democracy, or curing cancer, or ending world hunger, responding effectively to climate change, or peace for that matter, must concede computation can help produce better futures for more and more people too. The effect of my analysis of posthuman war should offer more than a simple denunciation of technology, since my argument in the larger framework is that there is no such thing as humanity without it. As realism demands, neither philosophy, nor science, nor good political practice can do what they purport to do without engaging in technological work, taking measurement, and crafting forms of organization of one kind or another.[24] There must be better and worse uses of technologies, which have existed since the time of the first human tools, about 3.3 million years ago, preceding the origin of *Homo sapiens* by a very long time.

Manuel DeLanda's book *War in the Age of Intelligent Machines* is also written from a perspective of a human–machine interaction.[25] Picking up from Félix Guattari, DeLanda uses the term "machinic" to describe this combination. He is not suggesting that because machines have become smarter in war, war used to be fought in unintelligent ways. Rather, "machinic" and human forms of intelligence overlap in the struggle between war and peace. We must consider different kinds of intelligence to explain the expansion of war-within-peace as it happens in posthuman war. The World Wide Web was originally conceived as a military application called ARPANET. It was an attempt to make information-sharing resilient to nuclear destruction, and to help its warriors continue to wage war even within the context of human extinction.[26] The expansion of communicative means without the human being per se makes what Shane Harris calls "the Military-Internet Complex" a serious concern.[27] But it also makes the "data rights declaration" by the United Nations—to "receive and impart information and ideas through any media regardless of frontiers"—all the more urgent to uphold.[28] To break down the divisions putting humanity at the center of Western modernity for the not-very-long period of about three hundred years "has unbounded possibility for good and for evil," as Wiener says in *Cybernetics* (27). Cybernetics promises an "unbounding" of the human being (27). This "unbounding" expands outward toward a rebounding of connections to other entities where otherness extends to numbers, and therefore, beyond the replication or inversion of oneself. Number rules, but this is neither good nor evil. It is simply expansive, with multiple futures attached.

THE LORDS OF THINGS AS THEY ARE

Wiener continues to ponder the good and evil applications of computational technology as he develops cybernetics in other parts of his book. He does this most directly in comments about the "Lords of Things as They Are." Wiener's Lords comprise a "very limited class of wealthy men" (161–62). The word "limit" in reference to "wealth" circumscribes an antagonistic relationship between use of scientific inquiry and the false equivalence of money. The Lords of Things as They Are seek to "constrict" the flow of data in order to maximize the flow of capital, and this form of "constriction" is perverse because it accumulates money at the expense of advancing knowledge. Wiener's Lords are not simply limited because they have too many "things." They are limited because they enforce a mutual relationship between the ownership of "things" and what we might call *thing*-ification: an idealist relation between humanity and the material world, where matter is reduced to requirements of the market. The Lords of Things as They Are are limited in time and space: too small as a class, and too restricted by commodified modes of classification. Animal and machine may share operating systems in the way cybernetics theory says they do, but such sharing does not for Wiener automatically hitch technological possibility to the abuses of power and wealth. The Lords of Things as They Are keep humanity and things in a state of stasis and isolation, assigning value to only the objects money can redeem. The relationship Wiener's Lords have to warfare can be reduced to profit gained from the means of military violence. Recalling Paul Virilio, we might also posit a connection between the Lords of Things as They Are and what he calls "information bombs."[29] In the manner of Lording, information bombs are damaging because they destroy the more useful discharges of thought and action necessary to neutralize them. When the Lords of Things go to war, to cite Wiener again, "everyone loses" (161–62; 159). This is a limit upon crossing where there is no return.

In order to keep track of that kind of loss, the machinic dynamics of twenty-first-century war should be unpacked further along the lines that Guattari defines them, denoting a "functional ensemble": man and machine recombine so as to produce new "powers of enunciation that are material, cognitive, affective, and social."[30] Guattari continues in *Chaosmosis*: The "take off [of machinic assemblages] is particularly clear with military technological innovations. They frequently punctuate long historical periods that they stamp with the seal of irreversibility, wiping out empires for the benefit of new geopolitical configurations" (40). The machinic assemblage is here conceived as a war machine composed "of virtual as much as constituted elements, without any notion of generic or species relation" (35). In these two passages, as with the DOD concept of computational systems as a relation of "many-to-many," virtual representation becomes compatible with the physical kind through innovative forms of mediation. Machinic ensembles override previous categorical unities, putting new categories of being to the purposes of posthuman war, and putting posthuman war within peace. As such, media innovation can work to erode standard forms of military division, dotting the lines between humanity, mediation, and matter. Guattari is thus careful to include both cognitive and material forms of agency in his theorization of the war machine. Foreshadowing Stiegler, thoughts as well as things have "enunciative functions," and their significance can be understood through new ways of tallying the numbers.[31]

Building from this premise, a mainstay of posthumanist theory focuses on how animate and inanimate matter exchange and are composed of information. As the physicist and quantum computing innovator David

Deutsch tells us: "the fabric of reality" is part of a "mathematical universe."[32] Similarly, according to Donna Haraway's more famous "cyborg semiology," technological codes share material qualities with the composition of the human being per se. Because codes are the stuff of life as well as matter, they can be recombined to create new species and join them to nonbiotic objects.[33] Here the human being becomes "an artifact of technology rather than its limit."[34] Prosthesis theory thus refers to a "new genesis" where the body–artifact interface opens up "hypertechnical territory."[35] Consistent with Wiener's original claims, a kind of Cybernetics 2.0 joins techne and bios together, as newly discovered data pathways are used to reassemble living with nonliving stuff. This reassemblage provides an ethically advantageous alternative to what Pramod K. Nayar condemns as "species humanism," inspiring a call for what he calls "critical posthumanism" in its place.[36] Nayar continues, the "exclusionary nature of systems of segregation, difference, purity, coherence and separation . . . is rejected in favor of mixing assemblages, assimilation, contamination, feedback loops, information exchange and mergers."[37] The adaptation of cybernetic vocabulary here is impressive and useful. However, there is no account for the multiple uses of posthuman mixing, its availability for both "good *and* evil," or as Wiener would opine, the war-within-peace.

The ethical promises of posthumanism share an affinity with Timothy Morton's definition of "the ecological thought." In addition, they reveal a contrast between Wiener's Lords and the possibilities for reality expansion.[38] Morton's work also helps clarify how number rules in postmodern war. His useful definition of "ecology" eschews a bifurcated relation between human beings and so-called nature, instead scaling up the interrelations between human and nonhuman entities. Humanity is not simply "posted" as the word posthuman might suggest. It is "quantized,"[39] further bringing "the ecological thought" in line with computational knowledge. For Morton, ecology designates our brush against infinity, again recalling what DOD systems theory calls the relations of many-to-many. The ecological thought thus evokes what Haraway calls a *sympoietic* way of thinking about animate and inanimate systems, rather than an *autopoietic* one. "Sympoiesis" is a word Haraway uses to designate open-ended systems, with no beginning and no end, save provisional endpoints leading to surprises you cannot evaluate in advance. "Autopoiesis," by contrast, designates unit-based thinking, where temporal and spatial boundaries are predictable and centrally controlled.[40]

There are important ways to use sympoietic and autopoietic system theory to introduce the broad features of posthuman war as reliant upon the DOD's essential concept of "many-to-many." We can say about posthuman war what Cary Wolfe says about systems theory in general: "Far from eluding or narratologically mastering the mutational processes [of joining human and non-human beings], it subjects itself to them."[41] The "it" here refers to systems theory in general. But it is applicable to a war-systems theory specifically, because it too is full of mutation without narrative mastery. What remains enigmatic about posthuman war is how the concept of the netcentric battlefield affirms the sympoietic flexible and open-ended approach, rather than the autopoietic goal of exclusion and hierarchical control. Since posthuman war does not seek an end to war, but instead pursues the weaponization of means expanding into zones of peace, "sympoiesis" is its strategic extension. If you are operating a Guattarian war machine, or if you share the DOD's data goals, then your job is to create hybrid identity assemblages with military violence built-in from the ground up.

What role does ecological thought play in its posthuman war application? Can we use its realist emphasis on nonhuman agency to further differentiate posthuman war from Clausewitz's duels? For Morton, it turns out that ecology is not very intelligible on its own, infinitely vast terms, more often impeding rather than critically advancing rational thought. Categories allude to more attributes (to use a term from Spinoza) than they appear to contain. "Genus" hides more than it shows. But what intelligence can we decipher from the hiding? In *The Ecological Thought*, Morton is attracted to the "perversely sublime," to "speculative enjoyment [and] fantasy" (22). He shares with other realist philosophers the rejection of Kant's privileging of subjectivity as the fulcrum of politics and knowledge, the categorical imperative where humanity is the end of law and reason, and where human beings should above all never be regarded as "means" (124–25). For the ecological thinker-cum-realist, Kant presents a fallacious and idealistic counterpoint of unity and categorical completeness. His human-centered unities are fallacious because they are made possible only by ignoring real things in their full multiplicity. Morton's emphasis on parts exceeding wholes (a tenet elaborated on by Alain Badiou vis-à-vis

Gregor Cantor's set theory) is why "ecological art" is akin to "mathematics" (105–6). What needs be further explored, taking off from Morton, is how the computational foundations of ecological thought are relevant for posthuman war. Here again, number rules.

REALISM AND POSTHUMAN WAR

The connection between ecology and computation is essential for surmising the role that realist epistemology plays in posthuman theory. Morton is sympathetic to the cybernetic proposition of human animals and machines exchanging information, the data interface between human and nonhuman matter. But a problem emerges for his epistemology because the progress of knowledge gives way to what he playfully calls the "super natural" (45). Morton's use of the word "super" retains the mathematical inflection of how physical reality exceeds what is too narrowly circumscribed as humanity's opposite, capital-N "Nature." Here too Kant's categorical imperative for humanity as the fulcrum of sovereignty and knowledge is dispersed within the larger ecological "mesh." The ecological thought is thus described as the "uncanny" (52), a form of occulted baroque abundance. Morton's "hyperobjects" (objects containing a multitude of other objects within them) thus go beyond the human being in both kind and ken. Entities are at best surmised in an "ambient" way (103). In this sense, which Morton also connects to computation (28), he concludes: "We really can't know who is at the junction of the mesh before we meet them" (40). "Totality" is a "scary thing" because we are saddled with an insurmountable problem of system complexity. "Mathematics" becomes an impediment to the advancement of knowledge, rather than the knowledge we need more of to know the world and act better. "Quantity humiliates," Morton says, overturns "scientistic" pretense in this numerically superordinated way: "Coherence is unavailable" because reality exists on a scale too big to know (56). Instead of knowledge, "sublimity" steps in: Keats mourning scientific intrusion on the beauty of a rainbow, not Newton's revelatory prism.

By contrast, Haraway is reluctant to define ecology in this quasi-Romantic way. She suggests the "always-too-much" *quality* of "ecologically *quantized*" reality does not detach human beings from better and worse forms of "response-ability." Haraway resists the romantic figure of a "bondless, lonely" thinker, anesthetized from scientific reason within "the Man-making gap theorized by Heidegger and his followers."[42] The association between object ontology and Heidegger aside, Haraway offers a key clarification if we are to understand how philosophical realism can be used to know, used to make, and maybe used to unmake, posthuman war. In the place of knowing things and acting better, like Haraway, in *The Ecological Thought*, Morton sees the scaling up of reality, mingling the human and nonhuman; but unlike Haraway, Morton's ecology leaves us with "negative difference . . . fissures . . . [and] the strange stranger" (40). Reality expansion ends in the aesthetic sublime. Here we might refer T. S. Eliot's definition of the real in "Four Quartets" as "the thing we cannot bear much of."

Nonetheless, Morton's ecological thought is an important introductory marker for the turn to realist philosophy, especially the re-rendering of reality by numerical means, in the following pages. Meillassoux, Badiou, Lenin, William James, and Bergson are deeply interested in the rule of numbers pertaining to mind, media, and matter. In reference to these realist philosophers alongside their idealist antagonists, Kant and Habermas, I want to follow Haraway instead of Morton on the primacy of "negative difference." I do not regard the problem of "too much" as an impediment to knowing more and acting better, thinking instead of computation as knowledge advancement through the use of new tools. Tracing how demography, anthropology, and neuroscience have become applications of posthuman war, I hope to offer something more useful than a militarized form of the ecological sublime. Luckily for my purposes, Morton leaves to others the work of developing the relationship between computation, realism, and the expansion of military violence. "War is environmental," he remarks in an especially inviting phrase (49); "the ecological thought" can also be like a "horror movie" (49). Morton's nod to cinema is an ingenious gesture, given the machinic (cinematic, visual, and virtual) technologies central to all three war disciplines within the focus of this book. But the question I want to address, which neither posthumanist theory nor philosophical realism has yet proposed, is how we live and die within the techno-ecological mesh. How much ontological difference is too much difference for human ontology to hold, or rather, can we scale up reality while allowing the ability to distinguish between better and worse operating systems (20)? Can we cordon off ecological thought from network-centric war without presuming to exit from systematic thinking, root and branch? Can we be both humbled by infinity and smarter than before the humbling occurred? Why negative difference without more expansive, and potentially affirming, forms of sameness? Can we have our cake of what Morton calls sublimity and eat our science too?

In a response to Morton, and taking a different position from Graham Harman's object-oriented ontology, Jane Bennett offers a way to have sublimity and the advancement of knowledge. Morton charges systems theory with being reductionist because it applies scientific models to the physical universe, which he claims leads to idealist versions of Kantian holism. Harman similarly claims "entities" can only be known insofar as they are "quite apart from any relations with or effects upon other entities in the world."[43] To say that "entities" are "quite apart" from worldly effects means they must also be unknowable. Objects withdraw from rational thought, in this version of object ontology, because they are too big, and therefore too complicated, for philosophy to know: numbers not only rule but also *rule out* epistemic advancement. Bennett challenges this theory of withdrawal and stakes out a more optimistic course within the context of philosophical realism.[44] Against the way hyperobjects (Morton's term) are said to resist "even a fractious-assemblage model," Bennett offers a keyword: "use."[45] For her, the rejection of Kantian holism should not bar human beings equipped with the right tools "from any attempt to engage, use, or know" objects.[46] The use-value of knowledge would not give up on the strange multiplicity Morton emphasizes; neither would humility within the mesh deny the possibility of coherent thought. Scaling up, scaling down, and scaling everywhere in between is necessary for the possibility of knowing more and acting better. Bennett writes: "Since every day, earthly experience . . . identifies some effects as coming from individual objects and some from larger systems (or, better put, from individuations within material configurations and from the complex assemblages in which they participate), why not aim for a theory that toggles between both kinds of magnitudes of 'unit'?"[47]

For Bennett, the general heading for this act of "toggling" is "system," as the title of her essay "System and Things" clearly suggests.[48] The concept of system is important as it helps explain how demography, anthropology, and neuroscience have become war disciplines by scaling up the objects of their inquiries: racial identities, relations of so-called culture, and the human brain. System is a useful concept for elaborating how humanity, things, and computation overlap within posthuman war. It is also one of the most important terms in the turn to war-as-data, as the DOD's "many-to-many" conceptualiztion of military violence.

SYSTEM OF SYSTEMS

The title of another DOD document, *Systems Engineering for System of Systems,* repeats the word "system" three times for emphasis, the ubiquity of system being precisely the point.[49] "System" here refers to the means by which many-to-many becomes sorted in to one, and how one represents new and more challenging categories of being. Along these lines, Clifford Siskin makes a key pronouncement about the generative capacity of systems as a "way of knowing [what is] really there," and a "mode of production."[50] System designates the deliberate use of data-sorting and the production of new identities and expanded associations between relatively autonomous human and nonhuman entities. In line with the network-centric approach of the Revolution in Military Affairs (RMA), system is a concept allowing us to see how military violence works beyond Clausewitzian duels. For DOD purposes, the system of systems concept throws into visual relief multidimensional sets of relationships, mixing animate and inanimate matter on an integrated computational grid and expanding the battlefield to all entities in all places. System becomes a way to enlarge the theater of war beyond traditional boundaries. How large? All the way large: identities, cultures, and brains. System of systems means that any exchange of information between any number of entities—as large as the planet and as small as the neuronal node—can be manipulated for the purposes of war.

The DOD document *Systems Engineering for System of Systems* defines system as an engineering concept. But it is an engineering concept open to military goals. The system of systems concept, or to use the disturbingly appropriate acronym, SOS, designates "full human and technical interoperability" (1). The word "full" here indicates the breakthrough in military technology to operationalize "the many" in the SOS sense: Posthuman war scales up military violence in ways reaching beyond its traditional forms of personal, epistemic, and political containment. SOS theory enables the "arrangement of systems that can result when independent useful systems are integrated into a larger system that delivers unique capabilities" (4); and further, SOS engineering is successfully applied when "a functionally, physically, and/or behaviorally related group of regularly interacting or interdependent elements form a unified whole with the desired effects under specific standards through combinations of ways and means to perform a set of tasks" (3). In these passages, as in the previous DOD documents, SOS theory overrides the friend–foe, home front or battlefront, war or peace, human or nonhuman kinds of war-related oppositions. But it also applies realist philosophical concepts to military objectives: identities, cultures, and human brains are equally made available for military functions. Who one is, to which group one belongs, and how one can know, are questions mattering greatly (and gravely) in the SOS way of making war.

The prospect of scaling up to produce new categories of belonging, as established by Morton's ecological thought, and emended by Bennett's pitch for useful knowledge, is also addressed within the SOS paradigm. SOS tactics, as we have seen, break down the distinction between living and nonliving entities, merging subjective behavior and physical reality according to an all-encompassing network-centric data grid. SOS reconstitutes reality virtually, as an assemblage of functional groups, and seeks to produce "new properties as a result of the grouping" (5). The words "new properties" as used here mean the following: whatever group might be located on the network, its unity must be provisional and fluid. The "systems engineer" must be able to follow trace elements in one system leading to "emergent identities" (v) linkable to another. Functional groups remain relatively invisible as long as you think, recalling Haraway's critique of autopoiesis, the SOS is closed. But for SOS theory, systems are open in the more dynamic, sympoietic way Haraway also describes. Functional assemblages of any kind can be technically altered to make them operationalizable in posthuman war. Through the scaled-up "management of material and non-material solutions" (v), SOS applications use "open systems and loose couplings" (23). Moreover, in language mirroring sympoietic theory, "open-system warfare" is attractive because it makes possible a form of military violence existing "without a central management authority" (5). In the absence of a single governing logic to system, in other words, the more computationally nuanced logic of the SOS can be found. What was formerly invisible because too massive is what SOS reboots as the new war matériel. What appears absent within one system's functional grouping becomes functional in unexpected ways once the engineer finds compatible relationships between different systems. The dynamics of system are "useful to the needs of the war fighter" (v) in this ad hoc and contingent way, as targets oscillate in and out of view within multiple spaces and in real time. SOS doctrine uses "complexity theory" accordingly: to solve "problems of large-scale, heterogeneous information technology-based systems" (9). But "heterogeneity" remains enigmatic in posthuman war because oppositional relations are that much more liable to expand and reverse.

Ten years after the DOD's appeal to system theory, another U.S. military research paper, *Systems Confrontation and System Destruction Warfare*, turns to the Chinese People's Liberation Army to overcome the linguistic insufficiencies of SOS's limited English vocabulary.[51] According to Jeffrey Engstrom's *Systems Confrontation and System Destruction Warfare: How the Chinese People's Liberation Army Seeks to Wage Modern Warfare*, and characteristic of both the *Terrorist Recognition Handbook* and the RMA's network-centric war concept, the focus here is on "opposing operational systems rather than merely opposing armies" (ix). Once more, we see rule number one of plurality (too many enemies to know), and rule number two of proximity (you may also be an enemy), very clearly in play. We also see the expected cybernetic elements of "informationized" human–machine contiguity: "Information systems" reach across, and reintegrate, a host of new subject–object assemblages—"entities both kinetic and nonkinetic"—and in turn make assemblages to function in new ways (11; iii). "Outer space, nonphysical cyberspace, and even psychological spaces" are conceived in common within the "electromagnetic domain" (12). But note too: when it comes to differentiating between subsystem components, "Chinese terms for system are found to be more precise than the English for the word system" (2). Among the Chinese terms preferred are "*tixi*-system" and "*xitong*-system." *Tixi*-system designates "large integrated systems," while *xitong*-system means "discrete systems" (2). According to *tixi*-oriented SOS theory, there is "no distinct or objective conceptual boundary that can be drawn between most systems *(xitong)* and what constitutes the system of systems *(tixi)*. . . . It is a matter of perspective" (3). The relationship between *tixi*- and *xitong*-systems promises larger-scale system integration in sympoietic—that is, nonholistic, open-ended, and nonhierarchized—ways.

Chinese SOS theory responds to Morton's critique of system-as-holism in two ways, making ecological thought useful for explaining a new kind of war logic: first, the parts of systems, or their subsystems, always outnumber the provisional unity presumed by any single given system. According to the *tixi* concept, systems are always both closed and open. When and where they open and close depends on both the function and scale of the systems; second, scaling up systems to render larger swaths of reality knowable is neither a sign of systems being reductive nor an invitation to replace knowledge with the Romantic sublime. On the contrary, if you cannot achieve the operation within the perspective of one system, it means you need to work within the recombination of larger system sets. As an instance of posthuman war, *tixi*-oriented SOS is temporally "fluid" and spatially "boundary-less." But the word "fluidity" as used by war planners should not be taken to mean human beings disappear in posthuman war (though extinction may be the fate of our species). As an entity subsumed within war, the human being is more significant to military interests than ever. Identities, cultures, and brains are reconceptualized not just by but also as part of the posthuman war machine. Morton's call for "the ecological thought [to] unground the human by forcing it back onto the ground" takes on a morbid twist, but with an open-ended afterlife: Humanity's grounding is connected with its coming annihilation, as in the enhanced ability to put each other in the ground.[52]

Another realist, Bruno Latour, is correct to link changes in "the states of nature" to changes in "states of war and peace."[53] His celebrated invitation in *We Have Never Been Modern* is to reconsider the opposing constitutions of nonliving matter and human subjectivity, to become "terrestrialized" the way Morton wanted us to be "grounded."[54] Among the countless divides we ascribe to the knowledge revolutions of the seventeenth and eighteenth centuries, the divide between the world of things (Latour calls this "nature") and the world of people (he calls this "culture") keeps us unaware of the "countless meanderings . . . practical arrangements . . . and networks that move between, within, and beyond the two."[55] Latour seeks to "rematerialize our existence," fulfilling, not simply refuting, the promises absent from most depictions of Western modernity.[56] He continues:

"This means reterritorializing humanity or, better, though the word does not exist, *re-terrestrializing*" it.[57] The root word "territory," as much as "terrestrial," connotes what Latour calls a form of "re-politicization." Thus "the democracy of things" alludes to supranational political organizations in addition to ecological ones.[58] In the examples of posthuman war presented in this book, my focus is not so much on the opposition, or the supersession of nations, or their replacement with a parliament taking seriously the agency of physical matter. Rather, the term "terrestrialization" connects to the changing states of war and peace in more specific ways. Here demography, anthropology, and neuroscience achieve new signifance as war disciplines. In all three practices, humanity becomes terrain within an expanded, highly mediated, and fully integrated battlespace.

A provocative phrase from the *Terrorist Recognition Handbook,* "identity infiltration," homes in on how the computational expansion of the ontological category works in posthuman war (198). In chapter 1 on war demography, identity infiltration highlights the tactical value of counting racial and ethnic identities in order to maximize diversity and categorical fluidity. Here I offer a discussion of the U.S. census, which since 2000 has used a check-all-that-applies system to tally self-enumeration.[59] But the juridical mandates prohibiting racial discrimination since the civil rights era are undermined by political legislation, paradoxically affirming, indeed legally mandating, greater racial sensitivity by the state. Through the affirmation of difference, and by scaling up racial division, the liberal state is freed from civil rights obligations. The numerical expansion of identity options works in exactly the way our most influential theorist of liberal democracy, Jürgen Habermas, wishes to forbid: civil society is no longer presumed to hold the state in check for the purposes of democratic representation. To appreciate how posthuman war challenges civil jurisprudence along these lines, I extend my discussion of the decivilianization of civil society in chapter 1 to Habermas's Enlightenment predecessor, Immanuel Kant. Here I examine not only their shared dream of "perpetual peace" but also its inversion, perpetual war. Accordingly, identity infiltration challenges liberal notions of the public sphere developed during the Enlightenment, portending a twisted version of Foucault's race war to come. As I will detail further in chapter 1, a reconception of racial identity as a functional ensemble of strategically constituted parts suggests that the so-called politics of recognition have mutated.

Communicative reason is no longer the benign ethical process whereby intersubjective agreement is reached between rational citizen-subjects. Rather, Kant's all-important "realm of ends" turns into a forbidden "realm of means." The technologies of mediation both he and Habermas affirm as essential for civil society (print based, politically neutral, intersubjective, and consensus minded) are supplanted by new tools of communication, no longer separating the means and ends of war. More to the point of "identity infiltration," Kant's categorical imperative, the emphasis on the human being as the fulcrum of knowledge and the basis for moral and legal rights, is expanded to the point of ontological collapse. Here we see the rise of postwhite national consciousness weaponized as an emerging U.S. white minority. The epoch of a so-called postwhite America rejects Kant's transcendental notion of humanity and is best explained in the realist terms espoused by Badiou. Badiou reconceptualizes ontology as a philosophical subdiscipline of mathematics, giving us a host of conceptual terms to unpack the computational foundations of posthuman war. Badiou's concept of the empire of number helps explain the difference, key to decivilianization, between Habermasian intersubjectivity and the quantitatively scaled-up theory of posthuman identity described by Quentin Meillassoux as ancestrality. In addition to discussing U.S. census politics in chapter 1, my foray into speculative realism in Badiou and Meillassoux challenges the relationship between human subjects as qualitative abstractions and their status as quantitatively calculated, combat-ready assemblages.

In Badiou's terms, the "essential numerosity of being" goes to the heart of the U.S. Army's Human Terrain System program I discuss in chapter 2, with emphasis on that geologically inflected middle term: "terrain." Here I elaborate on the *terrain*-ing of the human being by focusing on both old and new theories of counterinsurgency (COIN). Whereas classical COIN doctrine speaks of "the wind of revolution," or popular uprisings in terms of "the war of the flea," updated versions of guerrilla war move from a metaphoric explanation of network-centric violence to the physical battlefield. But the physical battlefield is also rendered in a virtual way. Here human and nonhuman entities entangle as informational entities.[60] In 2005–6, the

U.S. Army deployed its Human Terrain System (HTS) teams in Iraq and Afghanistan, embedding social scientists within combat zones for the collection of ethnographic data.[61] Given their attention to "human culture," HTS teams draw upon anthropological methods to win a war of "identity supremacy," a term theorized at length in the much acclaimed *Counterinsurgency Field Manual* (FM 3-24) of 2006. The American Anthropological Association's (AAA) oppositional response to the HTS program was in some ways expected, and partly paralleled the position the group took against the American war in Vietnam. But the curious part of the AAA's statement against the HTS program insists that no culture occupies "neutral positions." Thus in chapter 2 I compare the netcentric approach to human relationships in anthropology and posthuman war, given their common concept of culture as a nonneutral zone.

The anthropological desire to get away from a phenotypical theory of race rules out a notion of human difference based on so-called blood. In turn, cultural influences and communal upbringing require qualitative (social, moral, and character-based), rather than quantitative (computational, numerical, and physical) kinds of knowledge. This subjectivist thinking is especially apparent in the national character studies work, for example, as done by Margaret Mead and Ruth Benedict. Both figures worked in proximity to Allied war departments, and both were keen to distinguish a liberal democratic version of Allied anthropology against the racist thought espoused by the Axis powers of Germany, Italy, and Japan. The moral turn in Mead's work sought to promote the universal embrace of the human being per se, which must proceed according to a value-free conception of rights, freedom, and social development. But this neutral stance is what both the HTS program and the AAA's statement against the weaponization of so-called culture rejects. The nonneutrality clause in twenty-first-century COIN doctrine corresponds with the AAA's position on human identity and war. In combination, the two very different organizations (one military, the other civilian) recombine, revealing what Gregory Bateson calls a "schismogenic" dynamic of war. Bateson's interest in the connection between cybernetics and what he called "ecology of mind" is very much to the point of the change in military technologies driving posthuman war.[62]

Classic COIN doctrine was developed from the American Revolutionary War, through the genocide of Native Americans conceived as legal "foreigners," to the calculation-defying actions of the insurgent Vietcong. In its twenty-first-century application, COIN proceeds through the *terrain*-ing of culture as expressed in the HTS program. The turn to "cyber-enabled counterinsurgency" (CE-COIN) after HTS further emphasizes the data-dependent nature of posthuman forms of war. The military concept of culture-as-terrain does not jettison the human being per se. Rather, HTS mapping applications are further developed to provide more capacious ways to enumerate humanity than could be found in previous counting technologies. Walter Ong's theory of information exchange as a process of physical transmission stands in contrast to a notion of communication as immaterial, referential, and neutral. I use Ong's work in chapter 2 to put the question of communication in the context of geographical space. In contrast with Habermas's notion of communicative reason, information-as-data in Ong's sense is no longer an abstraction providing idealized categorical holisms apropos Habermas and Kant. By conceiving of information exchange as a physical process, both as means and matter, and by conceptualizing human beings as information entities, communal relations take on "specific material-semiotic properties."[63] These properties (with indefatigable emphasis on the plural) are no less real for their reliance on virtual media.

In such COIN initiatives as Project Maven, "Algorithmic Warfare Cross-functional Teams" find patterns on the battlefield unrecognizable to the human soldier but legible through computational means. Off-site planning stations removed from the battlefield channel information in real time from the point of enemy contact to the sites of forward command. At light speed, Green Cells map sewer lines, rivers, and human insurgent movements, both flattening and expanding the concept of terrain. The only requirement for this version of reality expansion is that the landscape be processed in virtual form. Ink and paper could simply not keep up with the scale and speed of Green Cell mapping techniques. Similarly, from "green data," "white Afghans" are made. In this curious phrase, "whiteness" designates civilians of whatever racial or ethnic identity who happen at that moment to be armed. In the language of the HTS program, "white situation awareness teams" are responsible for combining both the human and geological terrain. Here the coming white minority conjured by the U.S.

census politics described in chapter 1 moves in chapter 2 toward whiteness in the larger context of the Global War on Terror. In its further-reaching anthropological application, the enforcement of fluid demography depends on technological advances in computational media where identity is determined according to functional rather than formally determined categories of value. The whiteness of the white Afghan exists in a context where friend and foe are impossible to designate for certain, and where battles are won and lost at the level of data-ontology. If the numbers add up as green information, you can move friend and enemy forces in and out of the category of white. As the Afghan whitens, fails to become white, or travels in and out of whiteness, green data can be recorded, transmitted, stored, and manipulated in real time.

My reference to whiteness in its subjective (demographic) and communal (anthropological) applications moves in chapter 3 to military interests in human cognition (neuroscience). Here I transition from the coming U.S. white minority in chapter 1, and the white Afghan in chapter 2, to how identity infiltration is related to the brain's white matter. The materialization of whiteness in this distinctly computational context refers to the "white cabling that connects neurons in one region of the brain with those in other regions."[64] Like the coming U.S. white minority, and akin to the invention of whiteness in the Afghani sense, the study of the brain's white matter is media dependent. But according to neuroscientists, whiteness in the brain is also identifiable *as* media. Here the brain is conceived as a data-processing entity where the substance of cognition links humans and machines together in new ways. Until recently invisible to scientists and neglected as passive tissue, white matter proves to be both ubiquitous and essential for high-level human behavior. If the communication cables (axons) are insufficiently coated with enough white fatty substance (myelin), then cognitive functions, memory, and the expression of emotion are adversely affected. We could say then that whiteness in its Afghani application, as well as in the sense of myelination, no longer designates a fixed identity of any one kind. It is instead a fluid substance, media matter, in the brain's case literally fluid. Chemical and electrical flows are now reconfigured as war matériel. For demographic, anthropological, as well as neurological whiteness, breakthrough technologies in census mapping makes humanity interoperable with kinetic weapons systems: from plug-and-play to plug-and-shoot. Humanity per se is redefined as a host of connectors leading outward to further connections.

By attending to what brain researchers aptly call "the interior terrain," war neuroscience also depends on quantitative expertise. Here computers and computer screens offer unprecedented accuracy for human terrain mapping—in this case, neurological census-taking, but on a larger scale than demography or anthropology allow.[65] Indeed, military research into human cognition is keen to regard the brain itself as a computational entity. Focusing not on insurgent identities or cultures, but on the cell as biochemical circuit, neuronal census work aspires by name to formerly unthinkable scales of enumeration. Characteristic of the DOD's SOS concept, this scaling up of "enumeration" produces "finer categories," malleable for war purposes at the level of "human circuit plasticity."[66] Drone proponents speak enthusiastically of putting warheads *on* foreheads. But the shift from American postwhiteness to white Afghans and, finally, to white matter, also means putting warheads *in* foreheads. While the brain's "parts list contains thousands of millions of non-linear relations,"[67] human cognition becomes recognizable as code. Encoding life, the war machine operates in vivo, not so much as opposed to counterinsurgency but as the final step in identity infiltration. Here war becomes forensics, and in turn, military neuroscience turns cognition into stuff.

In war neuroscience, as in the census work performed in the undoing of civil society, or in counterinsurgency campaigns keeping track of who may or not be identified as "white," mapping is essential. Thus, computational biology puts image and matter together in virtual form. Levi R. Bryant's term "onto-cartography" identifies a technique of mapping common to "the ontology of machines and media."[68] Virtual reality and war neuroscience are intertwined in this way, moving the network-centric battlefield into the most complex inner recesses of the human body. In posthuman war, once again, matter, mind, and media converge. The "opportunities in biotechnology for future army applications," according to the Board of Army Science and Technology, begin with the manipulation of "biological things."[69] In military research on the brain, as in war demography and anthropology, images gain a certain reality-creating capacity. This is the capacity to construe material elements, like objects, bodies, and terrain, as coequal with immaterial ones, like subjects, cultures, and minds. As a theory of cognition, virtual reality also helps explain the difficulty of distinguishing between posthuman war's hard and soft applications: mind and matter, the event and our knowledge of the event, the actual and the phantasmal, the visible and the invisible. As I will suggest in my discussion of Henri Bergson, the brain is first a maker of images, and only after this, the knower of things.

Consistent with realist philosophy, chapter 3 draws on both Bergson and William James to advance beyond the question of Kant's idealism as presented in chapter 1. Kant's emphasis on human subjectivity is replaced in chapter 2 with "numerosity" (Badiou) and "ancestrality" (Meillassoux). From here, in chapter 3, I move ahead with a notion of the human being composed of what war researchers across the disciplines call "signaletic material." Under this heading, the primary research entity of the U.S. military, the Defense Advanced Research Projects Agency (DARPA), has initiated its "biocybernetics" program. The goal of this program is to develop war-oriented neurotechnology, allowing seamless interface between brain functions and kinetic weapons systems. To explain how this works, Bergson's theory of cognition as a process of "quantitative discreteness" is useful. His theory of mind foreshadows war neuroscience, conceiving the brain as a biophysical and mathematical organ, as well as a virtual-reality-generating machine. We are now at full circle, inverting the golden rule of Kant's antirealist philosophy: the human being exists not merely in the "realm of ends," as he insists, but also as a form of means.

Posthuman war planners are keen to combine military and cognitive systems by linking synaptic firing to firing weapons in the more traditional sense. This happens according to the SOS thinking already described. National security researchers, for example, currently ponder the use of fMRI technology for determining the distinction between combatant and noncombatant. The general point of chapter 3 is to show how the human brain can be connected to the battlefield. But the further-reaching implications of war neuroscience is that the brain itself can be turned into a combat zone. In the language of cultural mapping used in war demography and war anthropology, neurons are also targeted as populations. They are populations activated or neutralized by the application of electricity to stimulate activities, behaviors, and memories useful and commensurate with war. Biology is reconceived as part and parcel of a weapon system. This updated version of the smart bomb has resonance both with the virtual media being used to map the brain and, more enigmatically, with a conception of the brain itself as a data-processing entity whose firing mechanisms can be aimed and controlled. So-called neurological whiteness, in this curious sense of being-as-matter, plays a role in the creation of new ontological categories as well. As signaletic substance, white matter plays not only a quantifiably determined logistical role but also one of qualitative recombination.

I am intrigued by Sven Lindqvist's instructions to the reader on "how to read this book" in his introduction to A History of Bombing.[70] He writes: "This book is a labyrinth with twenty-two entrances and no exit" (prefatory page). The description is intriguing not because I want to initiate confusion in applying computation and realism to posthuman war, but because the confusion I hope to relieve resists an equation between the explanation of posthuman war and an easy "exit" from it. Neither the technologies nor the philosophies discussed in this book are inherently applicable as military violence. Rather, the new tools and concepts explained here should help us understand important changes in the conditions of both war and peace. Wiener's cybernetics are worth recalling once more by way of concluding this introductory chapter: "We [mathematicians] dreamed for years of an institution of independent scientists, working . . . not as subordinates of some great executive officer, but joined by the desire . . . to understand the region as a whole."[71] In Wiener's dreams, scientists will "confine [their] efforts to those fields that are most remote from war."[72] But we have to ask now if his "dream" has turned nightmarish, as he seemed to suspect it might. A look back at computational dreaming suggests the cybernetic legacy could not long escape the associations with violence haunting the human-animal-machine recombination from the start. As Wiener realized in advance, the destiny of the first posthuman science was to spread beyond its peaceful confinement. Similarly, from Deleuze: "I sometimes seem to hear a tragic note, at points where it's not clear where the 'war machine' is going."[73] It's hard to know where the war machine is going, I would only add, because it's going everywhere.

1. War Demography

When I was President of Texas A&M University, I used to wonder whether it was scarier to be responsible for a vast, global network of spies as I had been at the CIA—or be responsible for some 45,000 students between the ages of 18 and 25. Well, now I'm responsible for more than two million men and women in uniform—and all armed.

—Robert Gates

THE REVOLUTION IN MILITARY AFFAIRS

The epigram above is drawn from a conference presentation delivered by the U.S. secretary of defense in April 2008 called the "American Association of Universities," at the fiftieth anniversary of the *National Defense Higher Education Act.*[1] It was part of a session called National Security: What *New* Expertise Is Needed? (emphasis added). Before remarking on what the "expertise" means for culturally oriented disciplines in the humanities—and beyond them—I want to pause for a moment on the word "new." Here Gates is wearing the two hats of university president and security-intelligence leader. And as such, what he goes on to propose is "a stronger relationship . . . between the military and higher education" (1). Gates singles out "history, anthropology, sociology, and evolutionary psychology" (3) as disciplines uniquely designed to meet the challenges of a new kind of war. Gates worries, "In the public sphere there is often the view that we . . . , [and] the social sciences [and] humanities, in particular . . . , are

at loggerheads." But he then goes on to promote a far-reaching alliance between the disciplines mentioned and the military's interest in "the world's
ideological climate" (2). "We must embrace the eggheads," Gates continues, "even to the extent of including the liberal arts" (3).

In U.S. Marine Corps parlance, the hybrid term for embracing the
overlap between the battlefield and public sphere is: "deviled-eggheads,
because eggheads and Devil Dogs share plans."[2] Of course, war has always
been connected to the production of technical, scientific, and the liberal
arts. The U.S. Defense Department's deep financial support of early versions of the World Wide Web—like ARPANET, NPL Network Cyclades,
Merit Network, Tymnet, and Telnet, all developed as war tools in the
late 1960s—is the most dramatic testimony of a partnership between
war plans and the most important communication revolution since the
printing press. The kind of sharing Gates is after—the recipe for "deviled
eggheads"—hinges on the embrace of new media as a new kind of war.

Before Gates, U.S. Army Major General Robert H. Scales, a graduate
of West Point in 1966, first earned a doctorate in philosophy from Duke
University. And in line with Gates's connection between the humanities
and arms, West Point's own Behavioral Science Leadership Department
recommends rigorous study of "the relativistic perspective on values, customs, and religion," suggesting further, a "liberal college education may be
the best seedbed for developing this perspective."[3] This weaponized brand
of cultural relativism is consistent with the US Army's Diversity Roadmap,
which presents identity formation as consistent with soldiers in formations
more traditionally conceived: "Our long-term vision for human capital
derives strength from diverse cultures, attributes, experiences, and backgrounds. . . . The many characteristics and backgrounds of our own formations are now added to the toolboxes for understanding the populations
in which we may be deployed."[4] Here identity is mixed and multiplied,
made of moveable and malleable attributes. Cultural difference is both a
commodity for exchange and a military tool, an untapped means for the
maximization of force. To the extent that diversity is remapped not only
upon the battlefield, but more significantly, as an extension of the battle,
one could say humanity's internal divisions become loaded with new political significance. The affirmation of *many* differences (the emphasis is
Gates's) applies both to the army's own ranks and equally to the ranks of its

enemies. In this sense, identity becomes a kind of improvised contraption of war. Diversity writ across enemy lines operationalizes local differences and moves across the traditional friend/enemy distinction: on both sides of the battleline, the attributes of opposition become mixed. This remixing is consistent with the network-centric logic of counterinsurgency applied in contemporary war theory. It is also consistent with Gates's positive response to the proliferation of difference in the demographic sense, as in the coming of a U.S. white minority by 2040. Under Gates, the military champions a strong commitment to racial and ethnic difference, and offers its participants the unique access to what he calls "cultural IQ." The former defense secretary notes proudly, "[In 2014] African Americans make up 22 per cent of the US Army overall," compared to being "less than 13 per cent of recruiting age men and women."[5] In 2020, 43 percent of the 1.3 million soldiers making up the U.S. Armed Forces are people of color.[6]

Notably, overcoming the loggerhead between academic eggheads and military multiculturalism brings civil society closer to the zones of war. According to what has been called the Gates Doctrine, the public sphere is presented as already absorbed within an array of warlike political forces. This is what the executive governmental order requesting the study of ideological climate marks out. Gates wants to channel defense spending toward "new disciplinary collaborations," like the Minerva Project, centered on the study of "everyday . . . culture and people."[7] This is part of an overall strategy of "devoting more resources to . . . soft power . . . beyond guns and steel."[8] The existence of so-called culture wars thus runs from the earliest practices of humanities disciplines helping to enforce the expansion of empire to more recent debates over which traditions are conserved or displaced in the context of liberal arts education.[9] But what is different in the contemporary setting called forth by Gates is a host of epistemic, ontological, and media problems merging together such that the human sciences are effectively equated with a kind of espionage, as in the Gates epigram above. We might take at face value his reference to university students as would-be cultural combatants. But it is more difficult to surmise how diversity changes from being a civil rights matter to an extension of war operating on both sides of the friend–foe divide. As no one would argue, identity has been reconfigured by digital media. But here, too, it is less obvious how the multiplication of difference functions across communication practices and the latest forms of weaponry.

The RMA, ushered in after 9/11, and the adjustments to counterinsurgency strategy coming shortly after, uses the term "total-spectrum warfare," with emphasis on the expansion of military operations within all sectors of human (and nonhuman) organization.[10] Gates expresses this difference by embracing "culture and people" as "force multipliers," where force moves through the means of communication as a way to track and influence social mobility. Similarly, the RMA seeks to mobilize multiple sources of enmity according to a newly expanded logic determined by a planetary-sized network of conflictual grids. In the lexicon of the RMA, as in war "diversity," the keyword "multiple" signifies a turn to *quantitative* reasoning proposing to subsume the *qualitative* aspects of so-called cultural IQ. Human difference on the network does not mean reducing diversity to stasis—quite the contrary. Identity is seen as fluid, crossing borders, and inherently embedded in war zones. Moreover, war takes the place of civil modes of information exchange, which were once presumed to be divided from state violence or at least free to be for or against war in an external relation to it. Instead of big armies, the RMA focuses on small insurgent units, as the term *guerrilla* warfare suggests: small wars, surprise wars, wars of uncertain duration; wars without clear divisions between those who are targeted and those who are protected; and most enigmatically, war as a weaponizer of identity itself. The turn to cultural IQ in full-spectrum war presents more precise ways of measuring small differences and doing so on a massive scale: shades of diversity proliferate as the U.S. white majority recedes; small wars displace large ones, yet no one is beyond the combat zone.

Future wars, Gates continues, will be predicated on "new, more malignant forms of terrorism inspired by jihadist extremism." But note, network-centric warfare on this order also updates the state's military prerogative to address politico-religious wars, such as jihad, as well as "human rights"–oriented modes of planetary violence, such as "disease, poverty, climate change, failed states," and "ethnic strife."[11] This is an exceedingly vast—we must again say total—expansion of military violence, involving biology and economy, atmosphere and public sphere, all at once and without any one identity relation on the network being dominant over the others for any finite period of time. The Gates Doctrine insists that war "requires the diverse points of view ... of the countries we are dealing with."[12] Here, in the second use of the word "diversity," both the spaces of war (networks, not clear oppositions) and the temporality of military violence (no clear cause for war, no beginning, and no end) are effectively warped.

In a speech three months after the National Defense Higher Education Act anniversary, Gates calls for universities to "obviate the need for military intervention in the future."[13] This places war within an infinite temporal horizon. The "new reality ... for America, and for humanity," means "the lines separating war and peace ... have become more blurred,"[14] Gates continues. The "next 20 years," he says again, will integrate "cultural, social, and technological change," and will therefore require an updated and expanded military-security doctrine. "Civilian agencies"—particularly universities— "will have to learn to stretch outside their comfort zone."[15] So what is new, one could say—taking directly from the lexicon of the CIA—is a recasting of the citizen-subject as a sort of multiple agent. This multi-agency underwrites the logic of posthuman war, where cultural belonging takes place within a zone of defense that is also one of self-targeting. In the posthuman war zone, peaceful forms of social unity within the public sphere give way to a kind of fractal subjectivity: the citizen-as-solider-as-spy. As Gates makes clear, identity infiltration is fundamentally a technical issue, even over and above one of culture, the public, or identity in any peace-abiding notion of those terms. Here, Foucault's theoretical twisting of Clausewitz's famous aphorism gets twisted further still: not war as politics by other means (Clausewitz), nor politics as a means of war (Foucault).[16] Rather, we might say: human beings becoming means, means becoming everything, and everything becoming war.

The U.S. *National Security Strategy (NSS),* composed in the wake of September 11, 2001, helps further explain the concept of identity infiltration.[17] Consistent with the RMA and Gates Doctrine, the *NSS* jettisons traditional ontological, epistemic, and political oppositions, mixing these, and embracing the more capacious scales of difference advanced by diversity as a tool of war.[18] Civil society no longer marks what Paul Virilio calls "civilian-ization."[19] In the figurative language of the *NSS,* twenty-first-century wars will involve "shadowy networks ... that reach into every corner of the globe" (5). Thus connected, the network-centric dynamic of

mapping local cultures expands outward toward constantly shifting political nodes. By scaling up the network, war becomes in sense prepolitical,
recalling Gates, a worldwide condition of ordinary existence: no life outside war. The *NSS* thus portends the permanent condition of military violence, not just as *inter*-national war (between states) but also *intra*-national
war (within them). This kind of war—the Forever War, to borrow the title
of a 1970s science fiction novel—will be "different from any other war in
history."[20] "We are fighting on many fronts . . . against a particularly elusive enemy," the *NSS* reads. Future wars will be "seen . . . and unseen . . . ,
and will continue over an extended period of time" (9). The Forever War
is everywhere and nowhere, and within it, there is no decipherable "we"
against which to find a univocally opposite "them," no end to war spatially
or temporally. "We" are conceptually obliterated by the cunningly simple-
sounding term of the "many."

Under these conditions, the Forever War demands we overturn a
centuries-old political ideal where the individual, civil society, and the state
are harmoniously aligned, and political violence is exported in defense of
the home population. "All innocents" are vulnerable, the *NSS* insists, al-
luding to a state of war at the planetary scale. Accordingly, the "global war
on terror" (so-called GWOT) signals a time of war where the "enemy is
not a single political regime or person or religion or ideology" (7), but implicitly, a multifarious one. A close reading of U.S. security strategy conjoining "innocence" with "vulnerability" underscores a crucial point about
the importance of war diversity along these lines. Here the word "single"
escapes assimilation within the old categories of civilian "allness." Races
and nations are no longer divided absolutely, nor is the human being something conceived of on the other side of war. What the *NSS* and Gates are
rejecting is an entire tradition of politics where singularity is presumed to
be rendered homogenous, normative, and whole, through reasoned social
interaction and political representation in the form of liberal democracy.
The singularity embraced by network-centric war presents a theory of difference, where divisions are as massive as they are subject to change. To the
extent GWOT has a tangible enemy, and to the extent the enemy is shadowy and soft, military violence crosses all boundaries and becomes uncon-
ditional. This is not to undermine the significance of military hard power.
The militarization of civilian police forces—warlike SWAT training, grants

to police departments to acquire military-style gear, officer training in the "Warrior Mindset" and "the battle mind"—are well documented.[21] But in soft unlike in hard power, and symptomatic of the subsumption of the public sphere within posthuman war, the suspect and civilian recombine.

But note too that the occulted status of enemy-as-me provokes a "*common* calling" (1, emphasis added). This expanded notion of the common, in spite of—perhaps also, because of—war's reach beyond traditional lines of opposition, is significant because it is so strange. The tendency of contemporary war to exceed binary paradigms in shadowy ways is significant because it leads to some of the biggest questions facing liberal ideals in the twenty-first century: How can singularity also be normative? How can multiplicity both evoke and annihilate the traditional categories by which people come together with each other on a scale as big as the world? I will consider these questions in a more developed way below in consideration of war and civil society as theorized by Habermas and Kant. But because we are talking about scaling up diversity, about categories and oppositions, about the human being per se, and about new media tools, we are also trying to find an analytical vocabulary sufficient for helping us understand posthuman war as an event of the many. We are trying to find an epistemology sharing a kinship with political philosophy, past and present, but also a new conception of fully implicated in numbers.

To move in this direction, Badiou can help us rethink transcendental notions of humanity as unified and outside political violence by thinking through certain concepts taken from math. He notes, war creates a "fraternal . . . community outside the normal rules."[22] This is a far cry from linking the public sphere to intersubjective consensus apropos Habermas. "Fraternity outside the normal rules" suggests a form of togetherness beyond standard public-sphere activities—an intimate condition, blending civilian with military themes. For Badiou, the communal is found beyond Virilio's notion of civilian-ization, or at least it finds a secret logic for belonging there. The logic is secret because war is hidden within the fabric of consumerist ideals, and because the violence necessary to maintain a public sphere at peace with consumerism is erased by civility itself. The state's relation to the public sphere is to maintain the uninterrupted "protection of Western comfort,"[23] Badiou reminds us. But insofar as commonality evokes war's planetary reach, the secret is out. It is no longer possible, nor for the state desirable, to deny the public sphere's dependence on the violence it conceals. In line with Carl Schmitt (and Agamben), Badiou's remarks on how social homogeneity comes from the state's ability to designate friend from foe emanates from its monopoly on political violence. A further question, addressed below alongside other realist notions, would be how national forms of homogeneity disappear in the context of posthuman war, and how scaled-up war diversity comes to replace it. This is where Badiou crosses philosophy with mathematics.

"Number rules," Badiou writes, and we can extend this declaration to the most important actions of the posthuman war machine. "Count!" Badiou continues: "Who doubts this today? For under the current empire of number, it is not a question of thought but of reality."[24] The term "common" in the NSS document cited above foreshadows this dynamic of "the empire of number," which is at the center of the philosophical realism Badiou affirms. The empire of number reveals how new tools produce new entities: not only the technological means for remapping difference within former zones of peace but also the better measuring of demographic difference as permanent, if also transitory, targets. The peculiar call for war-in-common presented in the NSS documents brings new intensity to the Althusserian mode of interpellative hailing: the "Hey you!" of the police. This calls forth a form subjectivity connected both to being an obedient member of civil society and being implicated in a criminal act. Extended to posthuman war, and with Virilio in mind, we might expand this scene into the de-civilianization of civil society.

In his way, the empire of numbers scales up a capacity of recognition forbidding identity from assimilation within the public sphere because its numbers are too different and too large. Once identity goes beyond a certain level of diversity, once the contours of subjectivity change from being a consequence of shared reason to being available as an application of force, who one is exceeds representative publicness. Demographic espionage breaks identity free from the neutral category of the human being through a recognition of the spy as equal to (because always lurking within) the unwitting citizen-suspect. In this instance, war's so-called common call rebukes a centuries-old premise of modern government, challenged so effectively by Schmitt, where the liberal state reliably differentiates between enmity and the protection of citizens from violence: under the conditions of posthuman war, national unity is ramped up in the hyperactive form of nationalism. But this ramping up occurs because the nation is about to disappear.[25] Political representation no longer exists according to the well-worn difference between oneself and another. In cultural espionage, the *other*—like the terrorist, like the disappearing white majority, like a military force composed largely of minority populations—is already *us*. But the "us" in question is not simply "other." Its *othering* reshuffles former states of being into emerging categories not recognizable according to traditional forms of political and social recognition. Populations are multiplied, divided, and reconstituted according to data-intensive state security systems. These systems render unseen enemies visible everywhere and nowhere at once.

U.S. CENSUS POLITICS AND THE COMING WHITE MINORITY

In consideration of Badiou's command—"Count!"—we can consider recent changes in the U.S. census as commensurate with decivilianization and in accord with "the empire of numbers."[26] In the wake of the civil rights movements of the 1950s and '60s, and for the first time since the original, constitutionally mandated census of 1790, the 1970 census allowed individuals to self-select racial and ethnic identity as opposed to being observer-enumerated by a U.S. census official. This new mandate of racial self-recognition was presumed to accord with the five official race categories as established by the Office of Management and Budget (OMB). The official five, known as Statistical Policy Directive 15, once comprised the longest-standing set of state-recognized race categories in more than two hundred years. They include American Indian/Alaska Native, Asian/Pacific Islander, Hispanic, Black, and White, the last two with a Hispanic/Non-Hispanic designation providing an ethnic buffer amid the big polarity of black/white.[27] Although it does take some work to find a political logic for why these categories are being allowed to change in the twenty-first century (about which more below), it does not take a great deal of analytical work to put pressure on the OMB official five.[28] American Indian, for example, designates "persons having origins in any of the original peoples of North America," but excludes native Hawaiians, effectively rendering them Asian immigrants. For reasons hardly accounted for by increasing birth rates, according to the 1990 census, the American Indian population was up 255 percent since 1960. The Asian category overturns the former Japanese/Chinese distinction, which was operative during the internment of the former during World War II and which, up until 1943, served to underwrite anti-immigration statutes excluding the latter.[29] Today the Asian population contains such ostensibly different peoples as Samoans, Guamanians, Cambodians, Filipinos, and Laotians. Black currently mandates the one-drop rule of hypodescent, otherwise associated with the 1896 separate but equal doctrine known as Jim Crow. But since black technically contains all people of African heritage, it covers the entire color spectrum. So does white, which includes among its official members North Africans, Arabs and Jews, and all peoples from India and the Middle East. Hispanic, which would have been counted as part of a Mexican race in 1930, but not so in 1940, was until 1960 subsumed under whiteness. Given the status of ethnicity in 1970, Hispanic still mediates the purer, late-nineteenth-century, black/white racial opposition. Hispanic therefore can legally include blacks from the Dominican Republic, blond, fair-skinned, blue-eyed Argentineans, and Mexicans who would otherwise be Native Americans if they happened to be born on the north side of the Rio Grande after 1857.

The increasingly evident inadequacy of the OMB official five has given rise to a new and definitive voice on the scene of census enumeration: the lobby for multiracial state recognition. Multiracial activism has all but adopted the language of the civil rights movement but with the effect of undermining the state's previous civil rights obligations. Multiracial activists speak of undoing the OMB official five as a "strategy of resistance."[30] Actively supported by the right wing of the U.S. Congress, these activists are eager to "challenge the dichotomization of blackness and whiteness that originates in Eurocentric thinking."[31] Curiously, and without historical precedent, the U.S. government is happy enough to comply with these postliberal ramifications connected to the end of the nation's white majority. But this is true only to the extent whiteness is paradoxically secured by hindering civil rights jurisprudence. In November 1998, after four years of intensive study by thirty-plus federal agencies, after thousands of pages of congressional testimony, and after a host of demographic experiments and self-recognition analyses costing more than $100 million, the OMB issued its new guidelines on racial tabulations for the 2000 census and entered uncharted demographic territory. This new mandate allows an option to "mark one or more races" for the first time since the initial U.S. census in 1790.[32] Even limiting one's choice to just two combinations, this new law stands to increase the tabulation from 5 to 128 possibilities.[33]

Categorical speculation on this order is something the NAACP understandably finds disquieting.[34] As the NAACP is quick to point out, this recent fine-tuning of identity-based claims for civil rights effectively reverses half a century's worth of racial jurisprudence. The organization has good reasons for thinking in this way. They are reasons linking back to a civil rights struggle that took place within an earlier context of war. During World War I, NAACP president Morefield Storey likened the sacrifice of black infantry soldiers in the trenches to the struggle against racist policy and opinion at home. If they were to contribute toward remaking the world, a wave of civil rights activism following the Allied victory in 1918 ought to afford the possibility of "making America safe for America."[35] Over seven hundred thousand black Americans registered for Selective Service in 1917. The 369th United States infantry, known by German troops as "Hell Fighters," were the first to reach the Rhine River and provided the longest service of any regiment in a foreign army.[36] Truman got the message by 1948, and on June 26 of that year issued Executive Order 9981, stating it was "essential that there be maintained in the armed services of the United States the highest standards of democracy, with equality of treatment and opportunity for all those who serve in our country's defense."[37] But multiplying the OMB official five changes the civil society formula, making it harder to distinguish the citizen's right to racial protection. Multiracialism as government policy after the 2000 census opens up a way to undermine the form of civil rights–based juridical redress won in part by Morefield's soldiers. It does so by counting differently, multiplying, and therefore complicating racial identity to the point of categorical collapse.

There is a kind of autogenic political confusion introduced to racial belonging as prescribed by the new census politics. This is not only encouraged but also legislated by the state. (There is indeed a legal penalty for deliberately getting one's identity wrong.) On the one hand, there is the NAACP's option of retaining Jim Crow–type hypodescent. On the other hand, new modes of self-description produce a form of racial complexity threatening to release the state's civil rights obligations. To point out this paradox is to foreshadow a form of identity politics blanking identity out in the name of self-recognition. The self-legislating citizen-subject betrays the unity of the state. This paradox is apposite to a technique of government based on the disintegration of civil society in the name of its preservation. And this paradox is logically consistent with a political ethos where the nation is engaging in autogenic forms of war within formerly peaceable domains. To recall Defense Secretary Gates once more, "ethnicity" (as well as race) becomes what the *Terrorist Recognition Handbook* calls an "X-factor": an essentially agentless—as in collectively omnipresent but non-recognizable—reality under the conditions of culturally embedded military violence.[38]

In his much-anticipated book, *The Inclusion of the Other,* Habermas reconceives modern natural law along the public-sphere lines as "citizens coming together voluntarily to form a legal community of free and equal associates."[39] Here "intersubjectivity" is understood as "a theory of rights that requires a politics of recognition that *protects* the integrity of the individual in the life contexts in which his or her identity is formed" (113; emphasis added). "The juridified ethos of the nation-state," he continues, "cannot come into conflict with civil rights as long as the political legislation is oriented to constitutional principles, and this to the idea of actualizing basic rights" (137). But what else do the debates over multiracialism and the U.S. census reveal if not the conflict between the legal community of the juridically secured citizen and the individual freedom of individuals precisely to self-recognize? For Habermas, ideally, the emphasis on mutual recognition should become the basis for "post-national self-understanding" (119). This he has in common with Kant's incomplete presentation of cosmopolitan subjectivity as outlined in his theory of rights (about which more below). The "juridified ethos of the nation-state" realizes its full potential, Habermas suggests, in an "actually institutionalized cosmopolitan legal order."[40] "The political culture of the United States," he continues, "enables everyone to maintain two identities simultaneously, to be both a member and a stranger in his or her own land" (118).

But Habermas is both right and wrong. We will see better why he is right in a moment, since Homeland Security makes every citizen a stranger at home, to be sure. But as the new U.S. census politics reveal, intersubjectivity contains a severe contradiction intimating liberal democracy's demise in the name of securing democracy. In spite of the obligation of the state to calculate race on behalf of legally protected minorities, the expansion of numbers crisscrosses between old and new races, collapsing the unitary philosopheme of the so-called people. As we have seen in multiracial claims on federal jurisprudence, the "juridified ethos of the [U.S.] nation-state" not only "comes into conflict with civil rights" but also terminates the state's obligation to protect civil rights on legal grounds.

In a critique of Kant converging with Habermas on intersubjectivity, Quentin Meillassoux introduces the term "ancestrality." In *After Finitude,* he asks us here to rethink "the central notion of modern philosophy, [which is] that of *correlation*" (emphasis in original).[41] The term "correlation" addresses issues of both identity and the production of knowledge and does so in richly innovative ways. Meillassoux is concerned, like all modern philosophy, with the relationship between subjects (as *qualitative* abstractions), objects (as *quantitative* matter), and representation (as media *tools*). There are two problems with transcendental philosophy then, which the realist kind aims to overturn: (a) The so-called objective condition is only objective to the extent that subjects have conformed to the world of things "in a supposedly indifferent way" (5). Objectivity in the Kantian sense rests on the "correlation between thinking and being, and never to either term considered apart from the other" (5); and (b) by extension, the *non*-correlational relation between "thinking" and "being"—where the "things" thought about both exceed and permutate subjectivity—is rendered unavailable to either politics or knowledge. Meillassoux concludes: "One could say that up until Kant, one of the principal problems of philosophy was to *think substance,* while ever since Kant, it has consisted in trying to think correlation" (6, emphasis added).

The lasting inadequacy of modern philosophy as it is countered in the realist vein lies in how representation is theorized. It is theorized in the transcendental vein as the adequation or resemblance of an object presented in correspondence with a subject or unified grouping of subjects. Alternatively, ancestrality considers the status of an object as having an illusive (but also *allusive*) status. Thus Meillassoux offers the terms "archfossil" or "fossil matter" to indicate the temporal orientation of the real world as being beyond human-centered conceptions of time. Similarly, the spatial qualities of reality exist on larger-than-subjective capacities of scale. The fossil as such is both matter and life, or at least the evidence of past life in the form of its persistent mineral traces. But this does not mean vastness on this massive scale cannot be linked to a new epistemology or politics of rights. To the contrary, the qualities—more precisely, the *quantities*—of data composing the "arch-fossil" present a unique encounter with a real world, where the real is "anterior to the emergence of the human species" (10). Posthuman war works within this realist realm. This is clear in the way the state "Counts!" individuals to *discount* them: the citizen as both friend and foe. Further to the point, arch-matter is antagonistic to the merely qualitative forms of philosophical reduction, emphasizing newly emerging quantitative antagonisms in their place. In that sense, arch-matter is the form in which subjectivity becomes material. Through its materialization, subjectivity becomes the substance of war.

There are not only material elements in both the subject and the object, according to this argument, but also—and this is crucial—an inordinately large number of them. In fact, there are an infinite number of properties in this most capacious conception of the real. We move then from the problem of scale to the resolution of category. The presumption of understanding the world according to closed sets of genus lies behind the correlationist's epistemic and political errors. To get at this problem of genus, realism borrows the mathematician Gregor Cantor's set theorem, which informs the high level of contingency lurking within Badiou's command: "Count!" Badiou initiates his objection to Kant by calling for empirically measurable "pure multiplicities" in place of idealism's generic reductions of "properties." This way of ruling by numbers challenges modern philosophy's binary approach to knowledge, which divides subject from objects, and in so dividing, presumes the unity of subject to subject. In short, a category only maintains coherence if we accept it excludes the properties of other categories. But for Cantor, alternatively, whatever divisions you start out with—subject to subject, subject to object, animate to inanimate, and so on—genus is itself already divided by an array of sublimated properties you cannot account for in the original set. Sets are split one from the other, but more enigmatically, sets are split from within. (What else is the conundrum of multiracialism as a demographic crisis?) Moreover, the categories you may wish to use to account for the first set of non-accounted-for attributes will in turn be divided by more to come: A set is always larger than itself.[42] As Badiou says in *Number and Numbers*, "There is always an excess of parts over elements" (211).

But how does this excess become operational as a new means of politics-as-war, undermining modern subjectivity and mobilizing "the essential numerosity of being" (211) so as to initiate decivilianization? Insofar as correlationism must end with a form of categorical unity defined as the whole people—"the representation of parts to make possible the representation of the whole"—then in quite literal terms, reality scaled up to larger values must be either ignored or rendered into absence.[43] Badiou writes, "The numbers we manipulate are only a tiny deduction from the infinite profusion of Being in Numbers" (211). Identity is not reducible to the dual logic of being either oneself *or* another; and neither—on the larger scale of "pure multiplicity" (211)—is political organization simply a matter of relationships between friend as the opposite of foe. Correlation in a political sense means counting people into categories, as we have seen, apropos the mechanism of the census-taking. To the extent that individuals are identified by the state, on the one hand, and identify themselves in the same way, on the other, then correlation on this order may be said to exist. The law of correlation—at no time *weaker* than in posthuman war—maintains a boundary between military and civil operations, cordoning off citizenship from political violence. But governing citizens in a non-"correlational" way, where the citizen becomes soldier and spy, means I become a target or informer without warning, and even at the same time. This is what identity infiltration designates: The citizen is both her law-abiding persona and someone treated as if she is about to break the law. She is protected from, and subject to, new forms of violence, from all sides at once. Only counting, or counting *more* as counting *differently,* allows this to happen.

The curious turn after U.S. census 2000, commensurate with the evacuation of liberal jurisprudence in its name, presents a misfire in "generic procedure."[44] This procedure collapses under the weight of counting greater multitudes of identity groups, and is consistent with the alleged presence of the state's innumerable and proximate foes. Traditional categories of citizenship divide and are encouraged to reassemble, but they are encouraged to reassemble in politically retrograde ways. This marks a pernicious turn against intersubjectivity qua subject/object adequation. Citizenship no longer affords harmony with the so-called representative state. Instead, the state embraces a mercilessly *quantitative* political reality, adjusting traditional forms of measurements in order to produce a composite of numbers exceeding previous forms of *qualitative* value. Scaling up Cantorian sets not as categories but as multiples, and creating new patterns within and across these new multiplicities, is more easily accomplished in terms of the physical space implied by demographic mapping. But the implications of category complexity on this numerical order have important temporal dimensions as well.

Badiou refers to mathematics as an "eventual site" for the production of new knowledge, with the emphasis on the "event." This indicates a displacement of time as well as space, which also occurs through the addition of identities found lurking within the ones we take to be self-evidently there. The event, therefore, conforms to neither temporal succession nor linearity. It is, to use a term from the RMA, a network-centric time scheme connecting digitized realist epistemology to self-enumeration on a larger-than-usual scale. But the usual caveat: the event's non-recurrence does not mean it is without structure. It just means that when the event collapses old organizational forms by adding new items, it refutes subjectively *pre*-given ideas about categorical wholeness: Subjectivity finds itself "always coming in second place."[45] Subjectivity is an unreliable aftereffect of identity's multiple states. It retroactively encloses, or attempts to enclose, a moment in time where numbers have already overtaken it. This occurs through human memory by holding on to lost time subtractively. Here subtraction means canceling out the additions to thought that might help us calculate a "new present." You could say that the numerical value of genus works toward reality reduced to merely human ken, while war "diversity" expands reality with new "tools" and replaces "ordinary life" with military value. "The linear order of Numbers . . . is our way of traversing or investigating their being," Badiou writes, "but in reality, the structures are the consequences of our finite thought, of that which is legible in Number as pure multiplicity. They depose Number."[46] Keeping in mind the juridical lexicon signaling a new deposition of numbers, pure multiplicity fits with the perception of time espoused by a typical *Special Forces Advisor Guide* (TC 31–73). As a counterpart to the spatial revisions implicit in the Gates doctrine of cultural IQ, the *Guide* remarks: "Cultures may differ greatly in their perception of time," and it is therefore "useful to make the distinction between monochromic-time . . . and poly-chronic time cultures, where time"—like identity—"is viewed as fluid."[47] There is no subject/object adequation in the way genus is disrupted by the event (Cantor), no inter-subjectivity (Habermas) in the *Special Forces Guide*. There is a kind of geometry, or better, geochronology, implicitly capable of mapping multiple time-based contingencies. We might call this *realist* time, or just *real* time, a temporal analogue for the spatial conditions of posthuman war.

There are differences between Badiou and Meillassoux on temporality. These differences bear mention only to the extent that they advance the explanation of how posthuman war alters both space and time. Meillassoux rejects so-called aleatory materialism—the event that upsets what's already known—as being "already submitted to laws."[48] He further insists that "chances" are only meaningful to the extent that they are "not pre-contained in their precedents. . . . The present is never pregnant with the future."[49] In the sense of rejecting standard notions of linear causality between present and future, Meillassoux understands the concept of potential to be insufficiently speculative because the speculation remains unidirectional. It is determined, if weakly determined, by a state of already-existing affairs. Potential is therefore a remnant of the immanence fallacy, where the other is merely an inverted version of me. Only here, in the temporal rather than ontological sense, my present is too narrowly open to a set of futures more vast than I can conceive.

Recalling Cantorian doubling, and the doubling of doubling, and so on, genus is only effective to the extent that the future is determined by preceding categorical ends and expectations. If you have those expectations, you should be ready to see them fan out toward more complex connections. Categories collapse because they are, *eventually*, loaded with singularity: too many particulars. Therefore, at the level of real time, we remain in Badiou's pure numbers realm: There can be a simultaneity of events in real time that are precisely the same event but are also massively different, depending on what other events they may be determined to cause. So, in the dimension of space, there is no self and other, no subject and object; but there must be ways of dis- and reassembling every self and every other, just as we might open up the contingencies between yesterday, tomorrow, and today.

This is precisely how the U.S. census mobilizes Badiou's number rules in order to exceed the black/white binary. However, the limited kind of racial recognition once upheld by the liberal state expands outward toward

cynically illiberal ends. The future of race tends toward category mutation, which can serve civil or decivilianized ends. Badiou thinks carefully
about multiplicity, as we have seen, confronting ontology with the rule of
number. Accordingly, "the concept of quantity marks the impasse of ontology."[50] This impasse is a direct result of the refusal to correlate thought and
being, claimed by Meillassoux as the province of post-Kantian realism, and
used in cynically effective ways to disjoin the state from its former civil
rights responsibilities. Badiou defines the "situation . . . [as] any presented
multiplicity." The "multiple," he continues, "is always in the after-affect . . .
of the numerical inertia of the situation."[51]

Numerical inertia as such enables preemptive, permanent, and globally ubiquitous war. To do politics as war by other means, apropos the
twisting of Foucault's twisting of Clausewitz, you must get *in front of* Badiou's mathematical "aftereffect." Managing "numerosity" (Badiou's term)
at the level of the census is consistent with an autogenic war logic targeting the subject the moment subjectivity is called into being. Badiou's grim
fraternity is underwritten by what drone theorist Gregoire Chamayou calls
"necro-ethics": autogenic violence on the population—and within it—in
the name of winning the peace.[52] The enemy is targeted before the enemy
intends to strike. This preemptive-war logic is the temporal flipside to the
spatially collapsing liberal state. Identity infiltration turns subjectivity into
a numbers problem calculated at a level of variation not just *too* complex
but also *wrongly* complex for maintaining civil rights. This contradiction is born out of a curious new form of racial incalculability promoted
by the up scaling of generic recombination. In this way, the liberal state
is being re-instrumentalized to address a new kind of numbers problem
where civil society is displaced by rejuvenated forms of crypto-imperial
advance.[53]

In the face of civil society's apparent disintegration, Habermas exhorts
the global transference of intersubjective self-understanding, what he calls,
after Kant, cosmopolitanism. To appreciate the full significance of Habermas's claim, we must also return to Kant, not only to his dream of perpetual
peace but also to its inversion: the condition, spelled out by U.S. National
Security documents, of the Forever War. Habermas seeks to update human
rights on the grounds of reconsidering international law. He does not, of
course, take Kant's original investment in cosmopolitan peace at face value,

and provides an appropriately focused historical qualification by referencing the Westphalia Treaty of 1648.

In *The Inclusion of the Other*, Habermas rightly points out, Kant "had no inkling of the World Wars . . . [or] of ethnic and civil wars . . . , [and was instead] thinking of war between regimes and states . . . , [which] still allowed a distinction between combatants and the civilian population" (167). He then proceeds to correct Kant's historical blind spot by naming the means by which "a 'civilized' resolution of international conflict depends." According to Habermas's defense of constitutionally protected liberalism, the intermingling of "combatants and . . . population" (169) is supposedly staved off. The expanded liberal procedures missing in Kant should thus provide the "legally binding character of an institution analogous to a state constitution [read here: the United Nations]" (169). According to Habermas, Kant simply cannot think of a "constitutionally organized community of nations" orchestrated for perpetual cosmopolitan peace. This is because such an arrangement would be "scarcely reconcilable with [his] own soberly realistic descriptions of the politics of his time" (170).

But can we say more about what is absent in Kant—the impasse, according to Habermas, where Kant's realism (Habermas's unlikely term) could not be reconciled with his theory of constitutionally organized community? Habermas is careful to remark: "Kant is thinking here of spatially limited wars between individual states or alliances . . . , but not yet of anything like civil wars . . . , guerrilla warfare and the terror of bombardment."[54] Kant's notion of limited war is consistent with the "normative regulation of international law."[55] Can we produce the realist insight that Kant did not have, and read him, contra Habermas, from the position of what goes missing in the essay on perpetual peace, and can we do so in a time distinctly other than his own?

Given all we have said about subject/object adequation, and more squarely on Habermasian turf, and given the critique of intersubjectivity achieved through Badiou's different emphasis on numbers, how closely should we read the title of Habermas's key book—*The Time of the Other*—and in what sense can the other be included in the time of posthuman war? There is the hint of an answer to this question by way of a caveat Habermas offers when he is speaking of the "sober realities" Kant leaves out of his state-based differentiation between war and peace. This points toward a limit in Kant's consideration of *perpetual* peace, which, for Habermas, is unsatisfactory for a temporal reason: Kant is situated within a particular "politics of his time" disconnected from our own. The contradiction between multiplicity and being is writ here not in juridico-normative terms but in the way Badiou's rule of number makes time and space warp in relation to each other. Here the (universal) term "perpetual" and the (particular) phrase "his time" collide. Because of this collision, subject and other are no longer intersubjectively" engaged. Historical specificity cannot be "reconciled' with—and indeed disrupts—philosophical certainty, given the movement of time apropos the "event." Kantian philosophy delivers "nothing" in the interest of Habermas's "intersubjective" goals, scaled up beyond Kant's isolated moment. But Kant's philosophy delivers *precisely* "nothing," which is not only good enough, but logically appropriate to the disintegration of the human being per se within posthuman war. This "nothing" is everything: the inability of Kant's texts to explain contemporary realities of war mark what is absent from his thought—namely, the pressure of the many against the false coherence of the genus called humanity, and the subsumption of civility within political force. This is where we must add another feature of discussion to the antagonism between subjectivity and numbers.

In the *Rechtslehre,* or the *Doctrine of Right,* Kant lays out a philosophy of universality, humanity, reason, communication, and indeed, quantitative agency *versus* qualitative values. Here we might reveal the philosophy of posthuman war as idealism's inversion: not affording the promise perpetual peace but instead threatening permanent war. Kant defines "the supreme principle . . . of Ethics" adjoining one's own will with the application of the law, thereby producing a peaceful equilibrium between identity, the public sphere, and the state, as reasonable extensions of each other.[56] In the *Metaphysics of Morals,* he writes: "Act according to the maxim possessed of universal validity for all Intelligents, for that ought, when employing means to any end, so limit and condition any maxim that may be valid to oblige a law for every thinking subject" (40); and further, "Reason applies every maxim of will as universally legislative to every other will, and also to every action whereby it is affected" (150).

In this account, sovereignty is the legal equivalent of moral reason, which is why, for Kant, there is no logical bases upon which to find conflict between the state and its citizens, no civil war, no people's revolution, and certainly, no way for war to find its way into the public sphere.[57] Rather, a citizen's agency is expanded by a self-imposed willingness to grant the other the same rights as myself insofar as we both obey the procedures of reason. Habermas's other is thereby assimilated as equal to me. The state is merely the political expression of this preexisting ethical ideal. Further from Kant, "an imperative is therefore a rule making necessary a subjectively contingent action, and thereby representing the subject affected by it as one who must necessitate his actions to harmonize with the rule" (109). Here, the unitary concept of humanity is defined as: (a) "a thinking subject," (b) a subject whose thoughts are conveyable through certain "means" toward certain "ends," and (c) a subject who is willfully "obliged" to recognize the inclusion of "every other will." These are the three criteria of a peaceful republic, one where the individual is "universally legislative, himself [sic] subject to these laws, and belonging to it as its sovereign" (148). Self and other, society and state, are "harmonized" in this way.

On the surface, there is no otherness in Kant, or certainly no otherness in the strong sense initiated by posthuman war, where the self–other relation collapses alongside the friend–foe distinction, and the public sphere overlaps with the war zone. For Kant's polity to work, the citizen *only* thinks, and the state *only* represents. And as an "end," the action of the polity's will subjugates the "means" of representation. One's will is not only transparent to oneself but also determined by a "categorial imperative" restricted in advance by the other's will in "symmetry and fitness of means to end" (109). The other's will is equally communicable and transparent, and in the summation of all wills, the state only passively receives the totality of wills: from rational people-as-ends to political representation-as-means, and never the reverse. This is the logic of Kant's "teleology": "the exact proportionateness of means to ends," and "the totality of the multifarious, which taken together, do in the aggregate compose one thing" (468). By placing "ends" over "means," Kant preserves the goal of rational communication both to unify multiplicity and render it into generically recognizable categories of the purposes of enforcing the law. In this, he leaves to the side those within the state—and they are many—who do not participate in reasonable debate, and rules out any connection between war-making and representative means. The term "means" designates communication occurring outside political force insofar as it is a mode of information exchange used by rational human beings within the public sphere directed to the state and not by it. Accordingly, "means [must be regarded] as mere theoretic or technical principles, [or as]tools." Such tools remain exterior to human identity because they obey "mere technical rules (159)."

Technology is associated in Kant with the physical as opposed to the mental world, e.g., "how he who would like to eat bread has to construct a mill" (159). It is here where Kant gets into trouble with the likes of Lenin, for whom the following proposition was true precisely in the reverse: "Who so wills the end aimed at, wills also the means indispensably requisite for maintaining it ... [as in] willing an object as my own effect" (109). In the materialist response to Kant, the process works in reverse: objects determine subjectivity, and means condition their workers in ways not commensurate with universal but exclusively capitalist "wills." But here, too, the transcendental dualism typically found objectionable within classical German idealism reveals a more interesting technical foundation in the separation between force and means.[58] For Kant, "the merely technical" stands to the side of public reason, which merely observes and controls it (170). This paradigm places mind over the means of production, rather than the reverse, as Marxian thought would have it. Indeed, citizens are only citizens because they are "independent, [and] exempt from wants" (149). This makes the citizen coequal with property owners who have achieved a certain level of wealth. Property inequalities are perfectly acceptable in Kant because property inequity comes after human agency, and is the result of unwise choices allowable in a "free" society.[59] Propertyless women, workers, and slaves are not calculated within Kant's version of ideal universality. On empirical grounds, the ideal presents a universality of the few. Universality in "the realm of ends" (148), like Badiou's critique of the subject, also comes too late, retroactively, and incompletely. Kant's categorical imperative works as an ideal of genus referring to an abstract category, rather than as an empirical concept open to scaling up: category, in this sense, "admits no proof" (467), and "the realm of ends" subjugates the "merely technical" application of means.

Kant continues, "Technical principles [operate] by force of subjective stimulants [which resist] universal validity" (99). In the same way relations of "force" are presumed to be outside civil society, and in the same way human will subordinates "objects," technology comes to be regarded as reason's opposite: "instinct" (467), or the "effeminate surrender to sensitive excitement" (562). Technology is bodily, but like the body, it is mere means and not the ends of humanity as conditioned by the categorical imperative. Such "physical springs [serve] as a means of advancing an effective and rational benevolence" (636). But technical means are only benevolent if they are coordinated by the debate of rational and disinterested (propertied and male) human beings. Otherwise, "means and instruments" may contradict humanity (134). To repeat, identity is the opposite of tool in Kant and as its subordinate. He writes, "External things rest not on our will, but depend on nature, and have, as irrationals, a relative [read: nonuniversalizable] value" (134). Moreover, he defines a "thing" as "that which no event can be imputed as an action" (410). Kant's rejection of materialism, and the strict demarcation between human consciousness and things, is consistent with his separation between mind and body. But the more interesting division for understanding war technology is his separation between human beings and mere means. "Whereas an Intelligent is called a Person," Kant writes, "being by the constitution of his system distinguished as an end in himself . . . he may not be used as a mere mean" (134–35). Kant repeats this mandate several times: the citizen should "never employ himself or others as a mean, but always as an end in himself" (148); and: "The Intelligent himself may never be employed as a means" (159). Kant's "realm of ends" connects technology to the theory of right only insofar as it connects citizens to each other in a peaceful and politically neutral way.

But the categorical imperative according to Kant's realm of means expands and falls apart in posthuman war. This is because humanity is refashioned precisely as means, and because means not only subjugates humanity but also refashions the human species as a tool of war. Number subverts the social contract by scaling demography up. Elias Canetti critiques the subjectivism that Kant calls upon in resistance to this rule of number as having "a head without a world."[60] The long-standing criticism notes how humanity's capacity for freedom exists only insofar as the human being is detached from phenomenal reality.[61] But Kant's theory of the universal subject reveals a more interesting problem, which is how his form of universality forbids the persistence of empirical numbers. Kant escapes the problem of numerical excess by demoting technology (means, work, media) as a bodily matter subdued as much by reason as by law. This is exactly where communication becomes a posthuman war tool, a force recasting the mind/body split in a way precisely inverting idealism: posthuman war works in the realm not of ends but of means. As the term "mere means" is used by Kant, informational processes are inapplicable to forms of human contact as connecting with actual bodies. Here "mereness" relates to the subjugation of the material world by the transcendence of thought.

But here too, where things dominate so-called humanity, and force replaces public will, we can hear Badiou's command: "Count!" When Kant declares the superiority of ends over means, he is emphasizing the relative value of abstract mental processes over the real—if also occulted—power of multitudes. "The material and quantitative perfection is *one only* (for the total of the parts of anything is one whole); but of the . . . qualitative (formal) there may be many sorts in the *same* thing, and it is this last alone we here treat" (469, emphasis added). The word "qualitative" here serves the role of placing the end of human will over numbers, and thereby bounds the many to a category of "sameness." This logic runs parallel with Kant's association of quantitative operations as singular, as in "one only," and therefore nonassimilable by the processes of reason required by the liberal state. But, to quote Badiou at greater length, and contra Kant: "Numbers govern our conception of the political, with the currency—consensual, though it enfeebles every politics of the thinkable—of suffrage, of opinion polls, of the majority. . . . What counts—in the sense of what is valued—is that which is counted. . . . Political thought is numerical exegesis."[62] If Kant's philosophy is subjectivist it is because his theory of the subject resists the technologies necessary for realizing the hidden conflict between politics and numbers. For Badiou, by contrast, the growth of numbers "enfeebles every politics of the thinkable" by limiting politics to an impoverished, numerically underscaled version of the so-called majority. It is also in numbers where materiality lies, as in the masses of propertyless individuals excluded by Kant, as well as the way posthuman war reconfigures diversity as a matter of force. Badiou's insistence on numerical exegesis is useful for showing how war and numbers are converging, given new tools for refashioning the human being as a political concept. Indeed, where is there a more pronounced arrival of his "return of the numerical repressed"[63] than in a time of war where epistemology gives itself over to politics, and where everything knowable is also a target? The symmetry proposed by Kant between the means for achieving a peaceful society and the ends to which he too optimistically argues must determine those means. This ends-over-means argument becomes reversed by a logic of numbers. Accordingly, rights-based discourse is subsumed within war.

Badiou helps us explain the disintegration of juridical norms writ here as a kind of legal correlation between subjectivity and the necessity of government. Here the promises of intersubjectivity turn into mass forms of numerical dissensus. This dissensus results from a technology of counting where number is realized as a physical force. As Badiou states in *Being and Event,* the politics of "numerosity presents a 'quantitative' relation, or relation of power, between situation and the state. A situation that presents one-multiples" (288). Kant's rejection of materialist philosophy, his affirmation of the transcendence of reason, and his understanding of technology as mere "means" subjugated by human "ends" are tenants of his thought consistent with his reductive treatment of numbers. "The state," Badiou continues, "can only re-present parts or compositions of those multiples" (288). To follow the political thread all the way through, the idea called "the people" has less and less to do with who the people really are in their full multiplicity. The contract between citizen and state, the one Habermas reasserts as cosmopolitan politics at the international level, rests on a more universal application of the law.

But, as Badiou writes, "this [the citizen-subject's] distribution of obligation and authorization makes the one—which is not—into *law*" (21, emphasis in original). This is Cantorian set theory pushed into a political direction: Am I an individual related to others like me, or are there others whose difference exists on a scale I am not able to know? What happens when the state no longer recognizes me as one of its own? This is the application of state power ("law") toward the other who is me (the "one—which is not"). The tension between who I really am and the legal definition of me bears a retroactive time signature, where I am no longer known to myself before the state names me. In the same way, we can discuss the problem of time to show how the genre-defying actions of set multiples create a situation where the parts exceed the whole. We can see from here how the state's legal recognition of me is too late in recognizing the multiplicity of the people and is unable to suture this recognition to a liberal orientation of rights. Instead, multiplicity subverts liberalism, and undermines the war *versus* peace distinction. For Badiou, there is no original subject-social-state equation. Rather, he puts decivilianization in this enigmatic way: "The state cannot be the same multiple whose state it is" (289).

But are we not, under conditions of permanent and preemptive war, in a peculiar and new *state* of the state? Badiou is very careful to recognize multiplicity as the state's unthought, in the same way he wishes to connect philosophy with the rule of number. The state's unthought is the same for politics as philosophy. And here we must recall not just numbers in the sense of demographic categorical excess but also the socially defying implications of the mathematical absolute. In the case of posthuman war, we might simply say that the state is publicly activating a capacity for inhumanity it used to activate only on the sly. Badiou invites us to think so when he remarks that the state, through its categorical procedures, functions by excluding multiplicity at scales it must refuse to remain stable. But Badiou does not mince words on the empire of numbers in replacing the dream of consensual unity: "This radical not human can be *inhuman*," as well (28, emphasis added); and further, "Mathematics provides philosophy a weapon" (17). Badiou's "mathematical schema of the infinite" (20) thus designates an "as yet unnamable swarming under the concept of Number" (67), which applies to left-wing and right-wing insurgencies alike. To think about the empire of numbers at the level of posthuman war does not negate a political rendering of mathematics. Rather he suggests that "the infinite reservoir of Numbers belongs to an open future in which ontological forms of numericality will be investigated" (67).

Further toward an investigation of such open futures, Habermas refers to a complex section at the end of Kant's text on perpetual peace, the full title of which is: "Second Supplement: [A] Secret Article on Perpetual Peace."[64] The first item of significance concerns the concept of the supplement. As Kant's title indicates, we are again in the familiar territory of quantification, or occulted numbers—multitudes exceeding the generic norm. This explains the relationship between the disintegration of the liberal state and the proliferation of difference. The secret is Badiou's rule of number, a population gone massively wrong in the state's eyes, or more specifically, the struggle to correlate identity, sociability, and the law apropos Kant's categorical imperative. The second issue is the way Habermas wants to find peace as between supplemental numbers and political representation. Here he avoids a lengthy discussion of Kant's demotion of technology as mere means. Instead, Habermas insists on a harmonious alignment between identity, society, and the state. The media tool through which Habermas posits the Enlightenment ideals of civil society is stated more explicitly than Kant as the exceptional capacity of print media to produce the people's sense of themselves. Insisting on an uncrossable line between the state's monopolization of violence and the public sphere's autonomy from war presumes information exchange flows in a free and equal way. The civilianization of the individual is simultaneous with her *literary* integration within civil society. But Habermas is also clear: The unfinished project of Western modernity at its moment of inception did not assimilate the numbers of its subjects en masse. It did not achieve categorical wholeness in any empirically measurable sense.

The Enlightenment was of course full of popular contention, riots, and civil war, as any serious student of eighteenth-century history will know.[65] Eighteenth-century rioters make up a secret historical archive, identifiable supplements to the public-sphere idealism of Kant's day. Though not enough has been made of the persistence of such multitudes, Habermas does allude to print capitalism's inability to grant beneficial membership to those he calls the "masses." As universal as it sounds, the early modern concept of the human being per se was a distinctly minoritarian affair.[66] Common humanity was an idealized abstraction, one counterposed to the more nettlesome problem in Habermas's *Structural Transformation of the Public Sphere*, variously, of "the majority," "the common," the "lower sorts," "the masses," and so on (18). The population at large was in fact supplemental to the juridical and communicative norms of "the [historical] phase we call liberal" (77–70). Small numbers of individuals created the concept of universality in the modern sense, and they did so by carefully delineating the status of the subjects to be counted or not. Habermas has a telling way of putting this numerical sleight of hand. He writes: "The identification of 'property owner' with 'human being' as such" manifests a specific "ideology" (88). But also note that public-sphere ideals, however historically unrealized, are both "ideology and simultaneously *more than* ideology" (88, emphasis added). The persistent problem of the *more than* goes without satisfactory analysis in the rise and fall of civil society. In this sense, the quantitative reality antagonizing Kant's qualitative idealism both returns and is now seen as already there. Presuming capitalist social relations are predicated on peace and not war represents a certain—Habermas calls it "blissful"—moment that we may imagine but never really had.[67]

In making a special note of Kant's emphasis on the role of philosophy, Habermas endorses the state's consultation of the philosopher, which he takes to mean the "important role that Kant rightly accords publicity and the public sphere."[68] The role of thought is a characteristically intersubjective one. Thought is influenced neither by materiality nor by the multitude. Habermas is concerned about the historical inability of the public sphere to assimilate those outside its peripheries but leaves this well enough to the side. He thus remarks, "We may forgive his [Kant's] trust in the persuasive power of philosophy and the integrity of philosophers. . . . He [Kant] counted on the transparency of a surveyable public sphere shaped by literary means to open arguments."[69] But the question is not whether to grant Kant historical forgiveness. Instead, the task should be to look closely at liberal governmental reason during the moment of its apparent collapse. In this moment of collapse, we may better come to terms with the public sphere's never having actually existed. We can also see how its contradictions lead logically to posthuman war.

Recalling Habermas's theory of communicative reason, it is significant how he homes in on subjectively based protocols for knowledge and the technological means appropriate to achieving intersubjectivity. The technology Habermas depends on for philosophical effectiveness within the public sphere, and therefore outside the politics of state, is, as already noted, literary: intersubjectivity as mediated by newspapers, magazines, and the technology of the book is concurrent with the origins of the Western citizen. Print enabled, or seemed to enable, a certain way of denominating the world that subjugated quantitative reality.

In introducing the difference between Kant's historical moment and our own, Habermas also suggests that we consider temporality itself as central to the connection between identity, civil society, and numbers. A key element of the public sphere is the notion of "trans-temporal continuity."[70] Put simply, this means different individuals become sociable by sharing the same experience across different moments in time. Within the moment of social consensus, any aspect of individuality except those shared according to the limits set by rationality-as-literacy is excluded. This is why there is no room for affect in the public sphere, and why aesthetic experience must be relegated strictly to arts-based disciplines, made subordinate to the rational discourse of the critic, divided from science, and made adherent to the standards of taste. The shared experience of time is the fundamental basis of intersubjectivity in Habermas. This is one of the ways self-recognition conforms with the law. Moreover, time on this order is a literary achievement and, as the discipline was invented, a supremely qualitative one, not supposed to be a connected to "merely technical" quantitative practice.[71] How then do media technologies relate to the perpetuation of war?

THE GRAVEYARD OF THE HUMAN RACE

As Habermas recognizes, we cannot share Kant's time because his (synchronically) dominant historical conditions changed as history moved (diachronically) away from the Enlightenment into some other epoch we experience today. Habermas's appeal to forgive Kant his reliance on print technology opens to a different time than the Enlightenment one he describes, and therefore, to time *itself* as other than transcendent or continuous, in the Habermasian sense of those terms. But on closer reading, Kant's desire for a never-ending peace presents a disturbingly double-edged quality all its own. This must be part of the lesson of disjoining his time from ours.

We should recall that the title of Kant's "philosophical sketch" is drawn from a "satirical inscription" on a Dutch innkeeper's signboard, which places the phrase "perpetual peace . . . along with the picture of a graveyard."[72] And just as Kant evokes "men in general" for the Habermasian phrase the "human being as such," he also evokes what we can now say is a nonexistent and in the end a *satirically* stated correlationist balance between civil society and the state. On the one hand, there are the "heads of states . . . who can never have enough war"; on the other hand, there are the "philosophers who blissfully dream of perpetual peace" (93). The whole possibility of cosmopolitan peace Habermas wishes to endorse in his extension of Kant rises or falls on this opposition between politics on the one side and the transcendental philosopher's dream of peace on the other. This is again to put "ends" over "means." The production of knowledge must be regarded as "technical" rather than political, and the practices of civil society must be cordoned off from state intrusion. The point of granting to Kant the full weight of his inscription—the tongue-in-cheek equation between perpetual peace and death—is to suggest that the separation between communicative reason and state violence was already imagined by him to be impermanent. More enigmatically still, Kant's irreverent reference to the end of political conflict as also the end of humanity per se supplements philosophical reason with the self-evacuating rhetorical tactic of satire.

Kant begins and ends the essay with the Hobbesian premise of a "state of nature, which is . . . a state of war" (98). Recall that the most important form of agency in Hobbes is not the citizen, as some would believe, but the quantitative remainder of the multitude, what might call, after Badiou, the numerical repressed. Hobbes complicates his concept in the peaceable citizen by recognizing the persistent force of the *dissoluto multitudo.* In this way, "quantification" disturbs "qualification" (Kant's terms); and Habermas's human being per se is haunted by the masses. There is always a supplement to what is formally recognized as universal, we could say with Kant's secret supplement in mind. In Badiou's arithmetical philosophy, the category of the human being subtracts from the force of real numbers, which are left over in the case of contemporary counterinsurgency theory to become part and parcel of war. From within the constituted authority of the state (Hobbes's Latin term is *poetas*), and as a constitutive force of sheer power (again from Hobbes, *potentia*), multitudes existing outside the law can at any time overturn the state in the form of civil violence.[73] Kant's philosophy of Right rules out this kind of overturning. It does so according to the same logic of error Cantorian set theory locates as outside the unexpanded public sphere.

Accordingly, for Habermas's Kant, the most important mechanism for keeping the law active is to use genus—the juridical boundaries categorization provides—in order to win the game of abstract universality despite it being a minoritarian affair. To make the multitude presentable as citizens, Kant says: "The state of peace must be *formally* instituted," and further, "Peace can be formalized only in a *lawful* state" (98, emphasis in original).

Recall Kant's association for formal processes with qualitative over quantitative value. This foreshadows the ontological impasse Badiou notes in opposing numbers with human ontology: to be consistent, numbers require form. Therefore, my identity is consistent with the sovereign's (Hobbes calls this the *imperium*), and to seal this contract—literally, this suturing of myself to a suspension of time—I must lay down my arms unless they serve the interest of the state.

In Badiou's *Being and Event,* numbers supplement genus and can from there be connected to the quasi-Hobbesian conditions of posthuman war: contemporary COIN theory operationalizes the "numerosity of being" (279). The operationalization of numbers, as we have seen, persists in Badiou in a non-civil-society way. This can cut in humane and/or inhuman directions. To further detail how numbers trouble Kant's ideal of human universality, it is important to follow the "scale of measure" (283), or what Badiou calls "the instrumental value of the ordinal" (284). The term "ordinal" is here used in a taxonomic way relating to order in a series. The ordinal provides the bases for generic containment. But it also contains "a subjacent ontological signification"—the anti–civil rights implications of the new census math, for example—allowing both political and philosophical situations to be disconnected from normative morays. In that sense, the ordinal tries but cannot overcome the reality of Badiou's *onto-quantitative* impasse. Thus he offers a second way of thinking about sequence and sets, not the "ordinal" but instead the "cardinal." This term not only designates the provisional nature of numerical sets but also helps focus on the shared points between sets as a matter of category and scale. Recall that "a situation presents one-multiples" (288). Beyond that, a "state re-presents parts or compositions of those multiples" (288).

Since we are interested in states of being as foundational to the stability of juridical states, our focus must remain with Kant's on the connection of parts *formally* instituted as per the willful application by all citizens of the categorical imperative: all human beings. But for Kant in the context of his satirical turn, human beings find perpetual peace through the way number rules in posthuman war. Whereas Badiou's ordinal designates a situation of legibly sequential numbers as *one* set, the cardinal designates points of contact *between* sets. In that sense, cardinality presents points of constitutive power, "superior" (288) because the expanded sets are more dynamic in their associative capacity. The "cardinal is immense," Badiou writes, and "practically does not appear in 'working' mathematics" (287). To the extent that it is connectable to other points in any given set, the cardinal numbers act something like a node—a network—where different sets encounter features of each other, and from there, change the value of the sets themselves. Keeping in mind that identities are constructed, deconstructed, and reconstructed in the way we overlap with others in transindividual groups, the "cardinal" number can be seen as the mathematical equivalent of what I have been referring to by the term "identity infiltration." This form of transindividuality exceeds and—as we have seen in the case of census 2000—can be turned for and/or against an individual interest, or can lead to a form of individual interest the individual does not know she has. Hence Meillassoux's frustration with Kant's correlational imperative. But here it must also be admitted that Kant too supplements philosophical reason with the ironic nod that perpetual peace comes only when the human being per se is fatally threatened. Numerically speaking, the individual correlates neither to herself (because she is too many) nor to the ethical substance called civil society, which Kant wishes to keep within the socially harmonizing realm of ends. While on the one hand, Kant soberly requires the citizen to think within the realm of ends, on the other hand, he ironically acknowledges the *qualitative* limits of the public sphere. Put in Kantian terms, and with an eye on the Gates concept of diversity as an effective military tool, when perpetual peace gives way to the Everywhere War, the realm of ends gives way to the realm of means.

Summing up on Badiou, the concept of cardinal numbers positions large-scale differences in an alternative relationship with intersubjective forms of human ontology. At the level of so-called numerosity, liberal forms of communicative reason, and the premises of print-based media technologies, cannot hold on for long. For Badiou, "Categorical thought thus tends, at the heart of ontological presentation, to reduce quantity to quality."[74] In this way, "concepts" have a "quantitative identity," whose job it is, Badiou says in *Number and Numbers,* to mark a point of epistemic and social stability he calls "equinumeracy" (17). Back to Kant: the qualitative value of the human being supersedes its quantitative status if the categorical imperative is working fully. But it never does. "Concept," for which we can read "category," takes on what Badiou and the military researchers commonly call an "operational" function (17). "Starting from the concept," he continues, "we are able to pass through the object on condition that there is truth; that when we *compare concepts*, and that number names a set of concepts that they have in common, a property is made possible and defended by this comparison (equinumeracy)" (17).

What neoliberal census mathematicians have discovered, however, is something precisely the reverse of comparative commonality in this "equinumerical" sense. In the case of the U.S. census, the political function of dissensus—of "cardinal" multitudinous-ness—takes place according to a larger scale of numbers than ever before. By doing so in this *non*-"equinumerical" way, the state's new math hollows out civil rights as an extension of racial self-recognition. Kant's categorical imperative becomes a contradiction, because the state, civil society, and the individual are thrown into a *fatally* disharmonious state. This hollowing out of modern political subjectivity runs parallel with the perpetual peace essay's satirical evacuation of the "Kantian" philosophical position.

Recalling Hobbes's *dissoluto multitudo,* the game again is one of numbers. Global anthropologists note that the world's 184 independent nations contain more than 5,000 race or ethnic groups, with more than 12,000 diverse cultures.[75] Count five plus-*x* racial and ethnic categories in the U.S. census, 500 variants of cross-tribal affiliations in the cultural combat zones plotted out by the Gates Doctrine, and what kind of dilemma do we get? Badiou makes a crucial point underlying data applied at this scale, which he also puts, contra Habermas, at the core of philosophical speculation. To move beyond the reductive nature of "situation" (one number-group) and toward the more expansive problem of state (number-groups), Badiou remarks again, we must "construct what is essential for any thought of quantity: a scale of measure."[76] With this in mind, I should not ask of the state to simply recognize who I am, but instead, to calculate how many. The demarcation between friend and foe, and for purposes of posthuman war, the blurring of the citizen–combatant distinction, designates a numbers problem from the start.

Postulate number six of the first section of Kant's essay on perpetual peace says, "No state at war with another shall permit such acts of hostility as would make mutual confidence impossible during a future time of peace" (96). He refers to such "dishonorable stratagems" as "assassins or prisoners" and "spies" as the unthinkable practices of "soft" violence. I have referred to these practices variously as double agency, demographic espionage, identity infiltration, and just now, after Badiou, as mobilizing inconsistent multiples on behalf of posthuman war. Such stratagems are dishonorable because they breach the categorical imperative to equate the human being as such with property owning male citizens, the social compact writ *qualitatively* as sequential ordinal numbers. But more significant than a breach in the social contract, the appearance of combatants within the public sphere breaches a "trust in the attitude of the enemy" (96). Like the citizen, the enemy must be trusted to remain within legal limits, prohibiting the state to turn its enmity toward those it would protect. But this conceals a *quantitative* problem. In posthuman war, the sovereign sees the subject as an other—*many* others. Friends and foes recombine such that citizens become suspects independent of their will. In this way, cardinal numbers are enlisted to address a security breach already at work in the state's recognition of a population enmeshed within perpetual political violence. Once the breach of trust Kant refers to occurs, once the friend–foe distinction loosens up according to what Carl Schmitt calls "partisan warfare, perpetual peace flips into a "war of extermination." This, as Kant elaborates, means "a war on right itself, [which] would allow perpetual peace only on the vast graveyard of the human race" (96).[77] Badiou's reference to the humanity *and* inhumanity of the "Count!" command emerges here: The human being is coequal with the dead.

In referring to peace between human beings under the signpost of dead humanity, the satiric inscription of the innkeeper's dancing skeleton presents a kind of macabre hospitality. Here Kant returns us to Habermas's desire for philosophical integrity qua public-sphere transparency. But as I am arguing, civility in the liberal sense no longer triumphs over political force. The grim irony of a racial graveyard as the site of the eternal human respite suggests Kant may have surmised the devolution of the public sphere into Hobbes's war of each against all. In daring to pry open humanity's coffin as one way to conceive of perpetual peace, Kant reveals a form of war happening precisely *within* what he calls "right itself." The cancellation of rights is not a corruption of the liberal state. Rather, posthuman war enters civil society as an extension of liberal political logic in reverse direction: political violence is domestically activated rather than exported abroad. We must therefore make sense of the blockages in Kant's argument against what, like Habermas, he at other times idealizes apropos the categorical imperative: the arrival of a representative state, "capable of subsuming the general within the particular" (101). Here again, as is designated by the word "future," Kant reviles the presence of the spy lurking inside every citizen. Under the sign of perpetual peace (the literal sign), he denounces a politics to come where republican-oriented collective belonging turns into identity infiltration, and at the same time, peace converges with war.

To read Kant from the perspective of Badiou's inconsistent multiple, we must do more than explain his ambivalent embrace of republican potentiality. We must find something else to do with the lesson about how multitudes fail to add up to the universality Kant posited as the end-all of human relationships. To say again, the modern state depends upon tools of numerical measurement that are supposed to sustain ordinal numbers in a way both lawful and desired. Badiou calls attention to this as the finally reductive nature of public-sphere consensus. What bothers Kant in the failed struggle to achieve ordinality in this civil-society sense is still another—recalling the dance of skeletons, potentially fatal—rule of numbers problem. It is the problem of the supplement itself, the excess of cardinal numerical immensity gone beyond the categorical imperative by which Kant identifies the human being per se.

From a human rights–based theory of perpetual peace, Kant moves next to economics. He singles out the "credit system," where money "in its most dangerous form . . . provides a military fund . . . [that creates] ease in making war" (114). In this instance of how number rules, Kant puts two modes of calculation—political representation and the accumulation of wealth—at odds with each another. This contradiction between the logic of profit and republican logic is what keeps the philosopher-cum-public-sphere-deliberator from pronouncing with absolute certainty that war will remain separate from the practices of civil society. Amid this dilemma, Kant juxtaposes republican "approximation" to what he explores and then abruptly qualifies as "communal possession of the earth's surface." This is a move from idealistically correlating identity with the law toward more nettlesome questions about owning property and increasing wealth. The difficulty here lies in a problem Kant can only address indirectly through his morbid satire: the conflict—indeed a potentially fatal conflict—

between distributive justice and communicative reason. The inadequately calculated form of republican representation under the credit system leads to an ominous evacuation of the philosopher's own rhetorical position. Here life and death recombine. Kant refers to a specific kind of capitalism where money is not only the medium but the matter of exchange. Accordingly, the credit system obstructs his exploration of an earthly communalism outside the laws of private property. According to the abstract ideals of political modernity, individualism on the order of commercial society partners with perpetual peace. But Kant cannot think about earthly communalism outside the relations of personal property. Thus, here on (and under) Earth, peace is fatal. Social harmony becomes an extension of the grave. Put simply, credit ushers in a form of political violence poised to annihilate the entire human species.

To Kant's dilemma over the status of the Earth, we must also add the precarious state of philosophical knowledge. Kant's passages on the relationship between hosts and guests connect back to his satirical reference to the Dutch innkeeper's sign. Here philosophical thought is introduced as the would-be guarantor of perpetual peace. But this is only possible, as Kant says, through the "death of the human race." In this opening nod to the hospitality of the Dutch innkeeper, we must ask: Is the philosopher himself also a spy for the opposite side? Is he an assassin, a mortician, or simply a fellow rogue? Instead, the peaceable practice of free and open exchange within Habermas's celebrated lifeworld, philosophy is betrayed by a secret death wish for humanity. It is as if communicative reason has found itself overtaken by inhuman earthly agency, positioned between life and death with the enigma of having equally activated both.

As a mode of philosophical satire, the perpetual peace essay is poignant because it both promises *and* overturns the *Rechtslehre's* republican ideals. In further explorations of the Innkeeper's grim hospitality, Kant focuses on "the right of a stranger" (106). He proceeds from there to place certain limits around just how *strange* the "stranger" is. Kant's limits to the strange are the same limits he wants to bring to the limits of war—but as the satire suggests, finally cannot. The stranger and the friend stand in for distinguishing oneself from another. This is the same problem of distinguishing citizens from combatants, U.S. demographic norms from multiracial identity, and university graduates in the liberal arts from armed cultural combatants. In Kant's ambivalent theory of perpetual peace, as in the Gates doctrine, the RMA's emphasis on network-centric war, and the U.S. census after 2000, identity infiltration is at work. On the one hand, Kant entertains the word "stranger" to denote the "right to the earth's surface, which the human race has in common" (106). But paradoxically, commonness here means the estrangement of social belonging. It means the loss of civil rights as an extension of the law, and the exposure of the citizen to political surveillance, violence, and death, in the name of national security.

Kant places "communal possession of the earth's surface" into the "realm of peaceful mutual relations" (read: the public sphere), and therefore rules out the achievement of "democracy by means of violent revolution" (101). He draws a sharp distinction between the stranger's "right [as a] guest" and the stranger's "right of resort." He dismisses the right of resort in order to reject whatever pursuit of the commonness falls outside the public-sphere activities of "tolerance" and "social intercourse" (106). And in this ideal social arrangement, private property is preserved as essential: as a stranger it would be *too* strange for us to think about my things as also being yours. Thus, the harmonious equation between the human being per se and the humans being as private property owner is presented by Kant in a troubling way. His satire alludes to a problem where the connection between Earth and humanity is strained by two mutually exclusive prospects: one expands the concept of humanity to a point where communal possession is possible, the other exposes humanity to its coming extinction. To the extent civil society points the way toward perpetual peace, it does so like skeletons dancing above the Innkeeper's door. A conception of the Earth that necessitates eternal peace for the human species cuts in on social equilibrium midway through Kant's ironic dance of death.

Meillassoux's concept of ancestrality is relevant to the sinister exuberance inherent in the image of Kant's dancing bones. Here we can think further about replacing idealized notions of humanity with the fossilized trace of its coming extinction. As realist philosophy is right to emphasize, noncorrelational thoughts and deeds such as those above the Dutch Innkeeper's door by no means guarantee a happy ending for the human species; nor will property-driven conceptions of social organization lead its peaceable renewal. "The absolute," Meillassoux suggests, is a "reality that is thinkable as . . . the in-itself is indifferent to thought," and it is an "indifference which confers upon it the power to destroy me."[78] Further akin to the connection between philosophical realism and posthuman war, he continues: "The absolute is a menacing power . . . capable of destroying both things and worlds, of bringing about monstrous absurdities, yet also of never doing anything, or realizing every dream, but also every nightmare."[79]

Is this not also Kant's message on perpetual peace, with the ambivalent—both joyful and apocalyptic—image of dancing human skeletons at its core? In another enigmatic phrase, Kant suggests, the intersubjective effects of philosophy are dependent upon a "secret element."[80] This secret element is in line with double-agency, identity infiltration, and the peace-as-death satire haunting the more sober-minded public-sphere message of Kant's essay. What the *Rechtslehre* cannot speak save indirectly is the proximity between rights and war. Or at least, Kant's philosophy of Right cannot speak the relationship between humanity and the mineral traces of its extinction while retaining the rational integrity of the public sphere. Habermas's illusive goal of consensus-based intersubjectivity relies on the absolute separation of sociability from political force. But Kant seems to know better in his bones. He refers to the goal of perpetual peace as part of the "human inclination" to develop its relationships away from war as "the spirit of commerce" advances. And yet, under the war-enabling conditions of the "credit system," Kant remains ambivalent about the future of the human species. The likelihood of war or peace "is not sufficient to enable us to *prophesy* the future theoretically."[81] This is because the "spirit of commerce" can become equally the "spirit of death."

Perpetual peace thus designates both the unattainable approximation of the republican dream, on the one hand, and the extermination of the human being, on the other. This is the secret element of Kant's philosophy of Right, wrapped up in those doubly inflected words "perpetual" and "peace." There is also a secret element connected to the theme of time in Kant's essay. Temporality both intensifies what human beings have in common and gives commonality the incompatible values of sociability and death: the beginning of cosmopolitan peace is conceived at the same time Kant's skeletons seem to celebrate the annihilation of the polis. Common humanity thus denotes a secret element in the form of dancing death. But the secrecy is at least alluded to in Kant's comments on "hospitality." In a way that splits from Habermas, Kant retains a certain hesitancy to fully "trust in the persuasive power of philosophy."[82] This hesitancy presents its own rhetorical aporia in a text already aporic as a "Supplement." Habermas wants to "forgive" Kant for his historical blind spots. But if we agree to read Kant both in his time and ours—poly-chronologically—then the satiric quality of perpetual peace takes on a heavier significance that is only apparent in the epoch of posthuman war. Its significance does not have the logical function of rational persuasion but works according to the prophetic mode Kant is typically read as resisting. In this sense, prophetic discourse features in his supplemental sketch to signal the ubiquity of war within a form of society he could never have known in advance. Such a form of war is itself also prophetic in the sense that it is poly-chronological, preemptive, and permanent.

As I have argued regarding the Gates doctrine, the *National Security Strategy*, and the politics of the U.S. census, the republican dream of a three-way harmony among the individual, the public sphere, and the state, is apparently ended by the cruel awakening of posthuman war. This cruelty exists because the assault *on* civil rights is intensified according to a further securitization *of* them. The crossing over of state violence into civil society, and concurrently the subsumption of human rights discourse into the various spy-ops of culturally embedded war, gives Kant's ironic reference to perpetual peace a prophetic quality. Kant says, "The state will . . . invite their [i.e., the philosophers'] help *silently*, making a secret of it."[83] He thus places emphasis on both what is absent to transcendental thought—its numbers—as well what cannot be said about knowledge if you are bent on intersubjectivity as the ends of communicative reason. Identity retains its double agency no matter how much one wishes for consensus within the public sphere. Kant's perpetual peace essay not only reveals the necessity of doing philosophy as a matter of speaking publicly but also alludes to the secrets humanistic knowledge can no longer afford to retain. These secrets are the incapacity for the human being per se to peaceably accommodate its own numbers, the proximity between force and reason, and necessity of political violence for a property-based notion of rights.

RACE WAR

Next to Kant, Foucault's well-known work on security and war in the landmark lectures of 1975, *Society Must Be Defended*, makes useful advancements for understanding posthuman war. But note an unrecognized feature of this work within the many commentaries upon it: Foucault explicitly addresses how numbers complicate "the discourse of right" (52). Foucault sets his task, which is "to reveal revelations of domination, and allow them to assert themselves in their multiplicity, their differences, and their specificity . . . as the basis of multiple subjugations" (45). Whereas in the case of Hobbes's multitude a "binary structure runs through society" (51), a change occurs after Kant: no more Hobbesian war running "through society," but instead, a concept of sociability that monopolizes force as belonging solely to the state. In the case of Kant's harmonization between the individual, society, and the state as prescribed in the *Rechtslehre*, "For the first time the binary has been articulated a specific history, and groups, categories, and armies are opposed to one another" (51). This marks the Enlightenment age of political conflict where "the disequilibrium of forces manifest in war [are presumed to disappear] within 'civil peace'" (16). But Foucault regards war not in the strict terms of state versus state opposition, nor is political violence presumed to be detached from the maintenance of civil society. Here he offers language foreshadowing a different age of war, the Global War on Terror, "as exercised through networks" (*SMD* 29). What he calls "the statist unity of sovereignty" is internally divided, and the "subjugation [of individuals] within the social body" (27) no longer presumed. Indeed, the reversal of subjugation, in direct contrast to Kant's goal of perpetual peace, is what Foucault's history sets out to rediscover. "War is still going on," he writes, "with all its accidents and incidents of reversal" (51).

In his rediscovery of the Hobbes's multitude in the form of seventeenth-century Puritan insurgency, war is reconceived as "an instrument used in decentered camps" (61). War is defined as "strategic polyvalence" (76), an "emancipatory project not concerned with purity" (81), as Kant's categorical imperative on behalf of humanity might presume. Foucault continues, war retains its "aleatory element" at the "point of maximum tensions, as force relations laid bare" (50). Here he emphasizes war's "urgent call for rebellion or revolution" (88) not "neutralized by the reign of peace in civil society" (15). Foucault rediscovers Hobbes's multitude as an interior, though occulted, presence within the public sphere. The multitude exists as war, and as a form of Badiou's excessive numbers: "multiple bodies, forces, energies, matters, desires . . . , materially constituted as subjects"

(28). In the same sense, Foucault's own lectures take the form of broken histories, riddled with reversals and folds of enigma.

It is inarguably clear that the most important enigma in Foucault's lectures on war-within-peace is race. He presents a set of historical narratives seeking not only to establish the ways in which racial identification becomes something the state must protect but also to show how the citizen–sovereign relation mutates over time. Rights and their repression, demography, genocide, the whole modern juridical apparatus enforcing racism even while it sanctions rights-based racial resistance, all of these problems occur at the same moment rights make citizens governable.[84] In Kant, the human race becomes governable in the same way, presuming to overcome previous forms of social division where the property-owning bourgeoisie breaks away from aristocratic hierarchical fixity. This confrontation between property-as-commodity and property-as-bloodline works for Kant in the same way subjectivity transcends the overabundance of meaning in the objective world. By contrast, both race as a form of division and humanity as a form of inclusion function in partnership to exclude all that is greater than the human according to that oft-cited Foucauldian term "biopolitics" (81).

Two items under the biopolitics heading are important for moving the association between population and war beyond the domain not just of race but also the category of the human being per se. The term biopolitics—or at least the strictly *biological* part of it—may not be sufficient for covering the way in which race and military violence align in posthuman warfare. We might begin to access this by putting the emphasis not just on desire, as Foucault does when he references bodies, but more squarely on the issue of matter. The first issue to notice in Foucault on race is how the division between human and nonhuman entities tends toward abusive oversimplifications. Actually existing complexities crossing over between this division may exceed predetermined forms of racial belonging. Moreover, this may happen in ways that disrupt and reorient the unity of the state. Indeed, racial complexity may be encouraged or discouraged at different times to break the state/identity bond altogether. This is nothing more than to admit the Foucauldian strain beneath the fluid demography in the case of U.S. census of 2000 and the identity infiltration we have already unpacked in the Gates doctrine. The unraveling of identity—not simply by its repression, but by its recognition—signals enigmatic emphasis on diversity as a posthuman war tool. At one point in the lectures, Foucault uses—pauses over, backtracks, and then affirms—what he calls "race struggle or race war" (65). Here we see the excavation of a forgotten proximity between the concept of human population at the point that such a concept was invented—the unity it presumed, and the peace it was supposed to promise—and the practice of political violence. From the not-yet-fully modern position of the seventeenth-century peasant insurgent, Foucault's reclamation of Leveler race war overturns the concept of the human being as a unit of government.

Foucault thus starts his history of race war about a century before the earliest preoccupations with the idea that the state equals humanity rather than mere bloodline, apropos Kant's break with aristocratic rule. Race war in the Leveler sense starts before notions of race became sutured to the morphology of bodies, as in nineteenth- and early twentieth-century conceptions of state racism. The essentialist notions of race emerging at this more recent moment—identity as phenotype, skin color, the strict tally of bodily features as univocal race markers (81)—are rightfully odious for Foucault. But they are also examples of a race war quite different from the one he refers to. Against state racism on the older order, and in contrast with eighteenth-century disciplinary norms, he positions a more nuanced, and more politically nimble set of race politics presented by the revolutionary forces of impoverished English commoners. Their rally around the notion of race struggle was not at all a phenotypical matter, and it would be anachronistic to think about race in the sense of the categorization of bodies according to the essentialist categories of nineteenth-century biological norms.

Thus, for Foucault, during the first civil war of the modern era (the English Civil War of 1642–51), the fight for property in common against the dictates of primogeniture was race war in a special sense. Here we are faced with Kant's different notion of the common as (ambivalently) connected to Earth—a common right to resort on the earth, *and* paradoxically, to own the earth as private property to the exclusion of one's fellow human beings. Property here is what gives individuality its political legitimacy unless what humanity has in common is conceived as the extinction of the species. Insofar as the state is above all the protector of private property, by implication, the arch-protector of owners against the multitudes of propertyless subjects, Kant's Earth also functions to keep war outside the public sphere. But Foucault's praise of race war originates with revolutionary commoners claiming biblical authority to overturn private property. Thus Foucault associates war and race in a way not predicated on identity-bound governmental recognition, with humanity covering political force in the biopolitical sense. If the Diggers and Levelers were racially engaged at all, they were activated by a notion of identity predating Kantian juridical norms. They were bearers of a lost, but rediscoverable possibility where political force is laid bare as internal to social organization. As Foucault suggests, the Levelers are worth recovering because their kind of race war was *avant les outres*: the distinction between self and other, even as it operated in a so-called peaceful way, had nothing to do with the *othering* of races in the modern notion of that term. For the Levelers, during the regicide of the English monarch Charles I in 1649, war was racial, yes; but only to the extent that it posited a division between British national unity and an alternate definition of the human being as occupier of the earth-in-common. The earth itself was inseparable from seventeenth-century race war, as the names Diggers and Levelers overtly suggest.[85]

Preceding Hobbes's day, the Normans and the Saxons were at war in a racial way, too. But from the essentialist perspective adopted centuries later, such a race war would have looked oddly racial, as in white versus white form of racial violence. The stakes of Leveler race war against the security of property was also white-on-white racial violence. But it was racial before race became sutured to a later set of bodily norms. Only after race was invented in a modern way could citizen-subjects be given rights by the state at the expense of those who were the state's foes; and only then could racial difference become the unimpeded source of demographic knowledge.

It is at this point in Foucault's rediscovery of white-on-white racial violence where he adds the stunning remark: "It is the discourse of race war and race struggle . . . I am praising" (65). Now in what sense is it allowable, let alone desirable, to speak in praise of race war? In too many senses, it turns out. Foucault's praise of race war has meaning in the twenty-first century far exceeding what he might have imagined in 1975. To quote again from the lectures, for the Levelers: "Identification of people with monarch and nation with sovereign . . . no longer binds everything . . . but is replaced by a principle of heterogeneity" (69). He further comments, seventeenth-century race war is advanced as a "counter-history . . . of prophecy and promise," where "the bible is a weapon of poverty and insurrection" (71). As Foucault emphasizes, the weaponization of race in this very specific, seventeenth-century sense, was not, or not merely, biological. Nineteenth-century biopolitics created a certain categorical sensibility both producing the ability to recognize some bodies and erasing the ability to recognize others. The historian's task of reclaiming the divisions within the so-called white race is clear. What we lost to the march of time, you could say, was a history of "race war" without race in the purely biological sense. Thus the crescendo from Foucault's lecture and the full explanation of his praise of race war": The "homeland of the future," he says, "is populated by memories and ancestors whose genealogy has never before been written" (76).

To this already provocative claim, we must now add: "never before been written" until now. This is not to say race has never been weaponized. Any cursory study of the history of the human sciences will confront the link between race and war from the outset.[86] But Foucault's point about the premodern past of race war, and the one applicable to posthuman war in the twenty-first century, is this: Hobbes's multitude has been rediscovered and redeployed on all sides. Foucault's insurgencies and the battles of twenty-first-century race wars are coming to look a lot more like each other, and a lot less like either eighteenth-century "medico-normativity" or nineteenth-century "state racism" than is easy to surmise. Or does contemporary race war involve some unholy combination of all three earlier epochs? In some mutated sense, geology moves from background to foreground within the latest forms of race war, just, as we will see in the following chapter, the discourse of terrain moves back into the category of the human being. An exceedingly capacious logic of counting, dividing, and reassembling the elements of living—as well as nonliving—parts of the physical world lies at the core of contemporary COIN theory. The good fight attempted by the Levelers and Diggers against the enclosures of land has gone through a metamorphosis and turned into a demonic form of race war where humanity turns into both matter and means. In the latest incarnation of race war, the human species—as much as the earth—is what constitutes the battlefield.

In posthuman race war there is an oscillation between biology and geology, which means the term biopolitics may not be sufficient to capture the larger scale of race war, now a war within, as much as against, the human being per se. Now as in the past, race war hinges on three mutually reenforcing achievements: first, identifying new populations at levels of scale (then, the landless majority; now, the coming U.S. white minority); second, finding new divisions within old categories, putting new—and more capacious—ones in their place (then, no more monarchs, common land; now, decivilianized multiracial civility); and third, we see in Leveler race war a historical precedent for the re-weaponization of time (then, biblical prophecy; now, computationally determined preemptive first strikes, as well as permanent war).

Added to Foucault's account of race war, we must factor in better than he did the importance of media change. We must begin to rethink the state/identity relation not within Kant's realm of ends but according to the realm of means. For the Bible-reading Puritans, the specific media revolution of cheaply available printed material, and the expansion of English language literacy, helped give rise to the itinerant race warrior before the so-called republic of letters sought harmony between citizen and sovereign. The public sphere's essential relationship to writing, which runs through Habermas's public sphere and is essential to his communicative ethics, has been displaced by new capacities of computational intelligence (Enlightenment novels, criticism, and newspapers were computational too, just differently so). Today we compress and expand memory, take apart, add to, and put back together larger scales of data, in more expansive ways than ever before. This clearly complicates the kind of intersubjective print-based forms of consensus embraced by Habermas. The dynamics of contemporary war have folded back on the Leveler insurrection and allowed us to rediscover a lost connection between geology and biology. In this sense, counterinsurgency theory has moved through—and finally beyond—the soft power of mere culture. Or if war has not moved beyond culture, it has remixed cultural with material concerns. In twenty-first-century military strategy, communication is the very stuff of war matériel. Through technological innovation beyond print culture, a new war epistemology has enabled humanity and terrain to come together on a common network-centric battlefield. The full power of Kant's satiric evocation of perpetual peace as existing in the "graveyard of the human race" portends not just the intermingling of war and peace, friend and foe, life and death, and all the rest of the old military binaries, it also signals a profound change in the categorical imperative known as the human being. Security comes at the price of greater and greater invasions of war into civil society; and, after Kant, the salvation of humanity arrives only insofar as we get closer to realizing its potential extinction. In this sense, once again, the biological part of biopolitics turns from the merely human sense of life toward deploying political violence within the entirety of biotic and nonbiotic matter.

THE ALGORITHMIC UNCONSCIOUS

The conjoining of human and nonhuman entities brings us to Wolfgang Ernst's reworking of another key concept from Foucault: *archeology.* "Media archaeology," Ernst writes in *Digital Memory and the Archive,* "is generally associated with the rediscovery of cultural and technological layers of previous media."[87] How might we bring archeology in this sense to bear on the demographics of posthuman war? Even given Foucault's investment in the heterogeneity of human categories against state racism, and granted the radical vision of the Leveler insurrection, there are elements of Kantian subject–object correlation in the lectures still too close to the idea of the human species as more or less closed off from nonhuman entities. But in Ernst's revision of Foucault, "archaeology" calls for "media-historical narratives[,] . . . a chronological ordering of events . . . aim[ed] at formulating counter-histories to the dominant histories of technology and mass media" (55–56). Here we leave the realm of ends for an emphasis on means. Ernst's sense of producing data without narrative is developed beyond what we find in Foucault. The rewriting of dominant race history from the perspective of Leveler prophecy provides the temporal disruption, or the counter-history, both Foucault and Ernst wish to acknowledge in their work. But counter-history is not *counter* enough for Ernst because the concept of humanity is kept at the center of how we measure changes in the organization of the species by technical means over time.

Ernst's rethinking of an archeological approach to the human being—and his preference instead for a genealogy of media—is a useful way to reconnect with the quantitative processes associated with the displacement of purely human versions of collective agency. "The cultural inclination to give a sense to data through narrative structures is not easy for human subjectivity to overcome" (56). This sense of data without narrative—or at least, data without the narratives privileging human subjectivity as the arbiter of historical events—sets the stage for understanding how war technologies are beginning to give culture the primacy of matter. To "escape the discourse of cultural history," Ernst insists, we must turn from "mass media content" toward the agency of "non-discursive entities . . . , or machines" (56). The significance of this "contentless" approach to the study of knowledge-as-technology recalls the second Revolution in Military Affairs (RMA). The RMA, recall, insists upon an algorithmic revolution alongside the weaponization of culture. Ernst continues, the study of "non-discursive entities" on this order "belongs to a different temporal regime that, to be analyzed, requires alternative means of description" (56). The alternative means offered by Ernst—with an emphasis as much on the word "means" as on the word "alternative"—are expressly quantitative rather than qualitative ones. They are mathematically, rather than linguistically, driven communicative modes, precisely inverting Kant's peace-producing insistence to keep humanity vouchsafe within the realm of ends. "The cultural lifespan of a medium," he continues, "is not the same as its operational lifespan" (56). And for Ernst, the "operational lifespan" of human relations, or at least the version of human operations presumed to rise above technology and things, is well and truly over.

So what do we gain by following Ernst away from culture and toward numbers by way of machines? What can we learn about the remixing of culture, demography, and archaeology as ways to explain how identity becomes a new military tool? If we are to understand the stakes of posthuman war after subjectivity has been integrated into military technology, then we, too, will need better analytical options than the Kantian tradition provides. The clash between "human performativity and technological algorithmical operations" becomes less about the way humans fight against each other than how war is waged against (or indeed, within) the species overall. This is consistent with the analysis of a new kind of "race war": posthuman race war, focused in addition to culture on the "non-cultural dimensions of the technological regime" (61). Along these lines, Jussi Parikka refers to our "algorithmic unconscious" (2). Such a phrase closes the distance between thinking and media and places the two terms into the common theoretical register of "frequency ontology" (164). From Ernst's perspective, rather than marking some absolute time horizon of a catastrophic techno-future, electron tubes become our ancestral forbearers. Kant's dancing skeletons become Meillassoux's ancestral fossil traces, becoming in turn so many bytes and bits. In contrast to Kant's realm of ends, and beyond Foucault's historical liberation of human desire, Ernst's focus on media opens toward an epoch of war both old and new. He presents a way to think about communication linking machines and human beings where these entities recombine as "mathematicizable things" (59). The opportunity here is to close out an old premise about communication being subordinate to subjectivity, as well as subjectivity being subordinate to larger-scale realities of the physical world. The relationship between quantification, media, and humanity becoming matter is therefore the subject of the next chapter, which delves further into the topic of so-called human terrain.

2. War Anthropology

Almost any form of doing anthropology can ultimately be understood as a form of engagement. . . . The Commission suggests that a neutral position regarding engagement with security institutions may be non-existent in many situations. . . . To engage comes with risks of contributing to institutions with policies and practices one may oppose.

—American Anthropological Association

THE HUMAN TERRAIN SYSTEM PROGRAM

In 2005, the U.S. Army began its HTS program in Iraq and Afghanistan, embedding social scientists within combat zones for the "operationalization" of ethnographic data.[1] Shortly thereafter, the AAA delivered its objection, proclaiming the nonexistence of neutral anthropological knowledge.[2] In some ways, the AAA's latest opposition to using anthropology in war played out along the same ethical lines charted in the 1960s. The U.S. military's counterinsurgency program during the U.S. war in Vietnam was not embraced by most anthropologists in academic circles, unlike their close cooperation with the Department of Defense during World War II.[3] But in anthropology's more recent opposition to war, there is also a curious alignment. Anthropology is described by the war-inflected term "engagement." When Attorney General Robert F. Kennedy rejected a 1961 legal briefing citing the illegality of the Bay of Pigs invasion of Cuba, he stated:

"The neutrality laws are among the oldest laws in our statute books. But clearly, they were not designed for the kind of situation which exists in the world today."[4] This example of situational nonneutrality introduces a curious alignment between war and the opposition to war in the twenty-first century: neither the AAA in its condemnation of the HTS program nor the latest strategies within counterinsurgency theory wish to regard the study of human relationships as a neutral affair. Neutral anthropology is no longer an option. The AAA continues to maintain a strong commitment to a liberal ethics of racial plurality, identity rights, and, most crucially, given the historical association between anthropology and war, a separation between cultural analysis and political violence. So, on the one hand, it is firmly antiwar. But on the other hand, as the AAA clearly admits, politics and anthropology are embedded within each other within the context of political "risk." Regarding the analysis of social organization, there are no "neutral positions."[5]

Before examining twenty-first-century COIN doctrine as it stems from this premise, and before looking specifically at the use of anthropology in the HTS program, it is important to recall the prehistory of anthropology insofar as it is founded in war. Indeed, American democracy itself originates as a successful insurgency campaign. The linear designs of the battlefield so coveted by British gentleman gave way during the colonial revolution to a different tactical mode: British subjects picked a date, then formally turned against their former selves and became revolutionary Americans. Insurgent warfare of this kind was also present during the U.S. Civil War, despite the glorious charge of 23,746 men at Shiloh.[6] But how does the matter of insurgency relate to the study of culture as a nonneutral affair? What is the relationship between war within civil society and what the 1961 *Counterinsurgency Manual* calls "identity control"?[7] The notion of identity as a matter of force, and relationships between individuals regarded as military issues, intimates a long-standing relationship between anthropology and war.

For example, a "Statement on Problems of Anthropological Research and Ethics" approved by the AAA membership in 1967 refers to a 1948 resolution to "protect . . . the interest of the persons and communities studied."[8] This document was designed to counter previous AAA commitments to aid U.S. military engagement as exemplified by a 1941 AAA resolution:

"Be it resolved that the AAA places itself and its resources and the special-ized skills and knowledge of its members at the disposal of the country for the successful prosecution of the war."[9] This portrays a different way of positioning the discipline in relation to war than what the antiwar epigram suggests above. The more recent position adheres to Franz Boas's objection to President Wilson's "use of science as a cover for political spying" during World War I. Boas regarded any and all "political machinations" as anath-ema to anthropological integrity.[10] The idea of neutrality in anthropology as a neutral science was not objectionable to Boas. The "diplomat . . . sets patriotic devotion above common everyday decency," as he would have it. The "soldier, politician, and business man, merely accept the code of mo-rality to which modern society still conforms. Not so the scientist. The very essence of his life is the service of truth."[11] But where anthropology on the order of Boas sought to transcend the "imposition of external influ-ences"[12] in the service of truth, the nonneutrality position issued by the AAA in more recent times sees the discipline as riddled with partiality and conflict. Against the grain of science as an objective pursuit, the nonneutral anthropologist works along a political continuum never fully transcend-ing the power she may wish to oppose. The presumption used to be that culture could be studied "at a distance"—to use a well-known phrase from Margaret Mead—"where distance provided objectivity, to present the cul-ture and character of my own people in a way they would find meaningful and useful in meeting the harsh realities of war."[13] While Mead was explicit about the superiority of democratic values over Hitler's Germany during World War II, she also meant for anthropology to be separate from politi-cal force. Objectivity, in her use of the term, sits uncomfortably with the position of nonneutrality expressed by the AAA in the aftermath of 9/11.

In 1939, Mead founded the Committee for National Morale, which led in 1944 to a governmental counterpart in the Foreign Moral Analy-sis Division of the Office of War Information. On the one hand, this work signaled a way of connecting with war as part of the anthropological enter-prise; but on the other hand, as the adjoining of the terms "morale" (the Allied powers finding it) and "moral" (the Axis powers loss of it) suggests, power and politics forbid epistemic distance. Mead and her colleagues used culture as a way of getting away from the enemy's biologically based racist views. The Nazi emphasis on body and blood, not to mention soil, presented a kind of physical absolutism, *terrain*-ing humanity at appalling extremes, and rejecting the possibility of moral correction. In this way, the moral turn in Mead's work promotes a humanist embrace of universal values. But this, too, is less neutral than it sounds. The goal of universal emancipation as a moral goal also stresses the exceptionality of Western ethical ideals: identity, social cohesion, and assimilation are what Western idealism looks like. For Mead in the context of World War II, the good society meant "the supreme worth and moral responsibility of the individual human person" in generic terms.[14] In 1948, the British high commissioner, Sir Gerald Templer, invented the term "hearts and minds" during the twelve-year British war in Malaysia toward similar ends.[15] The phrase presents counterinsurgency with a similar move toward morale, emancipating humanity from the body, and inserting military conflict within a figurative subjective realm. So does the preamble to UNESCO's constitution, which reads: "War begins in the minds of men, and in the minds of men defenses of peace must be constructed."[16] Both instances underscore a post–World War II divide between the immaterial realm of culture and the physical one of biology. This divide no longer applies in posthuman military doctrine.

There are many historical examples of the use of culture as relevant to political force. Bronislaw Malinowski's 1929 essay, "Practical Anthropology," sought to define colonialist rule in the Trobriand Islands as "the control of Natives through the medium of their own organization."[17] Like the so-called Native problem for Malinowski, the practical part of the anthropological task in the American Indian context mixed an interest in cultural exchange with explicit military goals. But distinct from twenty-first-century war operations, the key word here is medium, not yet matter. Similarly, in a letter to Captain Meriwether Lewis, Thomas Jefferson separates "the purpose of commerce" and the desire to know "their [the Indians] language, traditions, and monuments, their moral and physical circumstances, and their laws, customs and dispositions."[18] Jefferson emphasized, "It will be useful for you [Captain Lewis] to acquire what knowledge you can of the state of morality, religion, and information among them; as it may enable those who may endeavor to civilize and instruct them to adapt their measures to the existing notions and practices of those on whom they operate" (2).[19] In this early example of war anthropology—and notwithstanding Jefferson's addition of physical circumstances—linguistics, history, and politics are given primary operational value. Such "knowledge, " Jefferson insists, should be "rendered to the war office" (2). Military historian Samuel Huntington is right to suggest the fight to expunge native populations from the North American continent made up "the golden age of military professionalism."[20] And the profession here included what we would today call cultural proficiency.

The advance of anthropological knowledge (a discipline unnamed until Kant) corresponds to the so-called civilizing aspirations of U.S. expansion in North America. Jefferson's command to treat the "natives in the most friendly and conciliatory manner possible" (2) was operative alongside military interests and, he hoped, would help promote the interests of national wealth. As Jefferson sought to extend white settlements as far as possible up the Missouri River, we can easily catalogue a characteristically American mixture of morality, war, and the production of humanistic knowledge. The principal author of the American Declaration of Independence invented a brand of cultural warfare before the term was used as such. His attention to what we now call ethnography triangulated Indian loyalties against French and Spanish trade interests, and on behalf of the advancement (and invention) of white American identity. With national-security-cum-epistemic interests in mind, Jefferson insisted to Lewis, "a copy of your journal, notes and observations of every kind, must be put into cipher," lest the colonial rivals of the United States "do injury if you are betrayed" (3).

The translation of cultural into military coding is interesting on its own terms from this historical vantage point. But the cipher remaining to decode in the epoch of posthuman war is not merely the long-standing connection between hard and soft forms of military violence. The study of culture and biology as war operations are historically intertwined, and their reintegration reveals new connections between media and physical matter. When John Wesley Powell founded the Bureau of American Ethnology in 1879, his understanding of history meant the progress of humanity through the same stages of development but at different rates of speed. This narrative of progress is alleged to have started from the position of nomadism and savagery, through a pastoral phase, then moving toward a point of civilization whose fulcrum was capitalist economic development. But interestingly, the American Indian, in Powell's view, was a problem external to the march of history. Ethnology posited that they had no history at all. This presented less a form of otherness to American identity than of otherness to the species of humanity at large. Viewed from this position, the Native American "must be either protected or destroyed."[21] Here the "natives" are both stranger to, and occupier of, their original home: the other as so other it is outside space and time. Paradoxically, the first peoples of the North American continent presented a "foreign policy" problem within territory they had occupied for millennia. As Powell advised further in his congressional testimony: "Nothing remains but to remove them [Native Americans] from the country, or let them stay in their present condition, to be finally extinguished by want, loathsome disease, and the dissent consequent upon incessant conflict with the white man."[22]

Clearly, in this monocultural rendition, the placing of the Native American as absolutely other than white afforded little hope of identities crossing safely and cooperatively into one expanded human category, as in Kant's dream of perpetual peace. Powell presented "the weaker race" in absolutist terms: disease or removal became ways to make the nation secure. But in Powell's testimony to Congress regarding Native American dissent, the connection between anthropology and political force worked differently than it does in posthuman war: in the nineteenth century, state racism essentialized human difference ("the weaker race") to scale difference down ("removal"). The attitude is different now. In twenty-first-century war anthropology, the relative fluidity of cultural difference is used to scale the concept of humanity up. The trouble with this example of upscaling is that the enlargement in scale exceeds the category of human being altogether. Anthropological relationships are cast in terms of political force, sharing the challenges and opportunities of other forms of the so-called battle terrain. In posthuman war, the *terrain*-ing of the human being goes still one step further: geology overtakes biology vis-à-vis a host of computational military tools. In the ever-expanding field of military operations, information becomes, life becomes, matter, and matter gets recruited on behalf of political violence.

The earlier form of state racism as exemplified by Powell is what Mead's later turn to culture attempted to displace. Even before Mead, as early as the 1930s, the AAA passed a resolution repudiating Nazi racial science.[23] Anthropology takes its shape as a cultural field from this moment on.

The horrific violations of Nazism crossed a sacred line within the discipline by adhering to a concept of nation beyond the liberal grounds of Boas's value tolerance, replacing it with physically based racial absolutes. This is the context in which Mead proclaims: "The deepest paradigm of cultural anthropology is the psychic unity of mankind."[24] Here mind and matter square off in opposition, and the body takes a subordinate relation to subjectivity in the transcendent sense: culture is presumed to counter biology as a character-building affair, exclusive of politics and violence. For Mead, physical differences are overcome by the inner workings of the mind.

Mead's appeal to the abstract unifier of the psyche differs squarely from Powell's weaponization of so-called native disease. Her ethnography addresses the mind, while his addresses physical matter. But both are interested in unity: Mead's is large and subjective (humanism); while Powell's is narrow and presents a kind of territorial blood line (white nationalism). "We *are* our culture,"[25] Mead writes, in a definitive statement. And further, from *The Study of Culture at a Distance*: "This manual is concerned with methods . . . for analyzing cultural regularities . . . inaccessible to direct observation." Notably, "this inaccessibility may be spatial because a state of war exists."[26] But Mead asks a more enigmatic question in placing war as a "spatial" obstruction corresponding to the temporal rift between the anthropologist and her subjects. "How many of us," she asks, "are available to use how many machines and weapons?" The question is meant to serve as a point of contrast between the technical applications of force and its opposite, the presumption: cultural is not technology, but simply "what we ourselves are."[27] Mead's contrast between technology and character study is evident throughout her work, where, as conveyed in *The Study of Culture at a Distance,* to get our proximity to the other correct we must work in expressly "analogical and not digital" ways (13). "Books, newspapers, periodicals, films, works of popular and fine art, diaries and letters [are] the sorts of materials with which the social historian has learned to deal" (3), Mead writes. In getting to who we are, however, an additional kind of knowledge is needed, that of the "anthropologist, who is accustomed to work without any documented time perspective" (3).

The space-time dynamic produced by print is contrasted with the immaterial expression of the oral interview. Here Mead shows a point of contrast between anthropological work and war technologies, while paradoxically setting the stage for the use of anthropology in war. She starts with the spatial obstruction of war. She then makes a categorically expansive move by adding to the discipline of social history anthropology's "benefit of interviews with living persons" (3). To get to the living, she must rebuke "the field of computing machines" (13) and "counting mechanisms" (51). Next, Mead makes a disciplinary move, offering a time solution only the anthropologist can bring to the social historian's spatial obstruction. By positioning media and disciplines the way she does, we get around war, or at least get a comfortable enough distance from it to adequately differentiate opposing cultures from our own. The distinction between the historian's print archive and the anthropologist's focus on oral narrative corresponds with a split between calculable forms of media and the immaterial kind. It is the unique prerogative of hearing over reading where national character supersedes biological concerns. But more significantly, Mead's insistence that "anthropologists . . . must continue to go to living primitive societies" throws into relief a tension between media and subjectivity: documented periods imply dead media, whereas a new mode of analysis achieved by oral interview provides just the right proximity between one culture and another. "Primitive cultures" move from being "spatially and temporally inaccessible" (5) to gaining a "cultural character structure" available for anthropological analysis (4, emphasis in original). Accordingly, "multiple clues" and "subliminal impressions" become knowledge about "individuals acting in nationally defined contexts" (24, emphasis in original). The "calculation" of the "multiple" remains subjective and not technological. Yet it also specifies "larger wholes" (24).

A 1943 publication for U.S. soldiers, the Pocket Guide to West Africa, serves as a case in point for the ascendancy of culture in war anthropology, but note how it lacks Mead's interest in knowing the enemy in the "living" way anthropologists "must go to primitive societies": "Everyone is entitled to his own prejudices, but it would be only sensible for those who have them to keep them under cover. None of us wants to aid Hitler."[28] In contrast to Mead, "going to" becomes staying away. The message here is, "We have our own racist ideas, which we are allowed to have as free individuals, yet we must also hide them lest they come to the enemy's aid." Culture may be "what we are," as Mead says; but if it works against victory in war, we can go "under cover." The Guide continues: "There are no men like Tarzan there [in West Africa]. You will find the rhythm, especially the off-beat, rather familiar—that is if you're a jitterbug" (3). "West Africans," the soldier would have continued reading, "are pretty much like people all over the world" (4). However, "it's best not to discuss political issues with them. When such subjects arise, listen and say nothing" (6–7). The message here is not: "Rid yourself of prejudicial thinking." Rather, you may retain the right to racism because it is *merely* thinking, your own, and your thoughts are immaterial differences allowed by individuals if they have them on the sly. Your prejudices can go under cover, or they are as significant as the latest dance craze, because they are cultural, not materially significant, differences.

But the body comes back in. The *Guide* says that if you are going to have sex with an African, "take a prophylaxis" (34). During World War II, condoms were standard issue items for U.S. soldiers. It is from this period we get the term "booby trap."[29] The prophylactic was as important for the fully prepared military kit as were rubber boots. Both kinds of barriers worked to keep foreign substances at bay. Revealingly for understanding the pervasive mind–body split in this epoch of war, contact with the West African in the form of sexual relations was safer than the swapping of racist thoughts. But both remained under cover. The distinction between mixing fluids in a sexual way parallels the soldier's instruction to keep your culture out of enemy hands, less the mix here abets the opposition. The key point lies in this triple separation: the cordoning off of one's physical fluids, the keeping of one's racist thoughts at bay, and the division between biology and cultural predisposition. This is consistent with the warning from a 1940s U.S. Marines *Small Wars Manual*. Here the focus is on a nexus between psychology and revolution acts: "When composed largely of mixed races . . . those [revolutionary] people present a special problem. This class is always difficult to govern, owing to the absence of a fixed character."[30] As with the sexual perils of the booby trap, the problem of mixedness unmoors categorical fixity. "Disaffection of the people" takes place as if a cultural prophylactic has been breached by a contagion called the revolutionary act.

Recalling Mead's opposition between culture and machines, the importance of media technology to understanding the mind–body split in the World War II era remains essential. In the 1940s, different kinds of media were applied to quantitative problems (mixed races) in order to produce qualitative solutions (psychic unity). "The fact is," the *Small Wars Manual* continues, "beside the great events of which history treats there are innumerable little facts of daily life which the casual observer may fail to see. Collectively, their power and volume may threaten the existence of the government" (20). Power is volume, we might also say. But it can be calibrated in either revealing or concealing ways, depending on the war fighter's tools. If your technology is verbal as in traditional war anthropology, you will be "listening to peasants relate a story, whether under oath or not, to give a bit of information." But be forewarned, "It may appear that they are tricky liars because they do not tell a coherent story" (24). To the extent that stories are on the side of the revolutionary, the best hope for the counterinsurgent is not to suss out the little bit you can. Find a way to discover both the right kind and right amount of information: volume threatens a coherent mapping of the battlefield because stories go in multiple directions. New tools are needed to make sense of human relations at this level of informational excess, and they are just the tools of calculation Mead rebukes in the interest of national character over physical difference. In the 1940s, the storyteller qua combatant (who may lie) gains advantage over the listener qua uniformed invader (who may be deceived). Thus, the crises of racial numbers behind the insurgent's deceit are twofold: first, numbers "mix"; and second, they exceed presumedly coherent categories. One of the ways to fix the mix is to put oral narrative into print. The desire to move from orality to the printed page is based on achieving more reliable modes of sorting, essentially a precomputational data processing maneuver: "innumerable bits of insurgent information" may be put in the form of worthwhile targets once we get things on publishable record. Here we must expand Mead's reference to war technology as merely mapping the battlefield and dropping ordinance. The war machine ought to also include the practice of ethnographic writing itself.

Before the wide use of radio in the decades after the 1940s, print media provided the best way to communicate Western propaganda, and was the primary means for transmitting military code. With an interest in addressing the Malay and Tamil peoples, the British Information Services Department circulated newspapers in both the local vernacular and Chinese. In 1951 alone, more than five million copies of such weekly publications circulated, the *Farmers' News* being the most widely printed. The number is significant given that the areas of British-Malayan interest had a population of not more than five million at the time. There were even reading rooms and information centers installed in rural villages.[31] These areas were particularly sensitive regarding racial matters, played out in terms of Chinese communists versus Malayan police. Moreover, against the multitudes of five hundred thousand jungle-dwelling rubber tappers, there were only about ten thousand police officers available to address civil disturbances. Insofar as these disturbances took place along racial lines—as the pluralized name of opposing forces, "Malayan *Races* Liberation Army" (emphasis added), explicitly expressed—the civilizing promise of literacy was presumed to construct defenses not of bodies but within "hearts and minds."[32] When Templer coined this well-known phrase, he was referring to the pacifying effects of print media, specifically. Indeed, the division between print and oral cultures paralleled the division between civilized and primitive societies in their march-or-die progress toward Western ideals. The question of print was in turn central to human progress as defined by the discipline itself: "In social anthropology," S. F. Nagel wrote in 1953, "we attempt to extend our knowledge of man and society to 'primitive' communities, 'simpler peoples,' or 'preliterate societies.'"[33]

Military occupation and the human sciences worked according to a common form of "rationalization," to use a word linking managerial thinking and knowledge production in Malinowski.[34] "The culture of the native," he continues, "compensates for the loss of his military power." This compensation, if administered correctly, enhances the security of indirect rule. But here, again, we have a great divide in anthropology over which forms of contact enabled colonial rule and which forms might challenge it. The administrator's problem lies in "obviating those situations in which an illiterate chief has to deal with an educated clerk. . . . The object is to create in a Native authority a devoted and dependable ally who is always in harmony with European requirements." More important for emphasizing the connection between anthropology, media, and war, native disharmony was not only politically dangerous it was also "sociologically unsound."[35] It was sociologically unsound because the so-called Native's power operated according to an anti-European conception of social relationships. In the Native's conception, humanity was presumed to be organized such that the objective goals of political force were immanent to the conditions of human (and indeed, nonhuman) organization. As such, the loss of Native military power is consistent with the goals of cultural anthropology in its imperial guise. In the context of war, anthropological divisions line up accordingly in this way: (*[subject + literate]* = *civilized*) = *harmonized*; and then there is the opposite: (*[object + oral]* = *native*) = *disharmonized*. But in point of historical fact, subjects and objects, minds and matter, civility and de-civilization cross over time and again.

As the turn back to bodies and matter in twenty-first-century war anthropology indicates, culture never thoroughly shed its material connections, though these connections have remained under historical cover. The appearance of the HTS program in the Global War on Terror thus signifies less a break in previous versions of counterinsurgency theory than a lost connection with older ways of bringing human bodies into war. The well-known war strategist John Nagl repeats a now-standard account of U.S. defeat in Vietnam as failing to value the full application of counterinsurgency theory.[36] The doctrinal shift leading to the RMA in the first decade of the 2000s—network-centric conflict within societies and populations, friends and foes incoherently aligned—was nothing more than a renaissance in the golden age of imperialist war. War strategists mourn the "lack of appreciation of military theory and military strategy [on irregular war], which led to the exhaustion of the army against a secondary guerrilla force in Vietnam."[37] They mourn the rise of the Weinberger Doctrine, which allowed for an emphasis on conventional war, with its clearly demarcated front lines, and flat oppositions led by actually existing states.

What war planners mourn as the incomplete application of COIN in Vietnam should elicit a message about the significance of network-centric military operations in the twenty-first century. In the earlier historical moment, data outpaced the tools by which the insurgent bodies and stories could be captured in a stable way. In the postwar years, new technologies promise fuller and faster results. This is not to say that bodies did not count in Vietnam. The infamous "body count" was not only appallingly high but also regularly documented. It was lopsided in terms of civilian dead as well. Between the early 1960s until the main U.S. withdrawal in 1973, approximately 58,000 of the three million Americans sent to Vietnam were killed, and around 300,000 wounded. Even by conservative counts, some

400,000 civilians were killed, 900,000 wounded, and 6.4 million turned into refugees. The total number of war dead has been put at 1.5 to 3.5 million people.[38] Because land mines continue to exact a heavy toll in Southeast Asia to this day, war casualties are still being tallied. Clearly, numbers mattered in the earlier version of COIN. But they did not win the war.

The counting of the dead in the U.S. war in Vietnam would be of less interest and, to be sure, less challenging, than the enumeration of the living in the war to follow, as if there were something about the newer technologies allowing the weaponizing of life itself. The Strategic Hamlet Program failed in Vietnam because of bad census counting. Attempts to make the numbers legible failed because they did not remain legible long enough to create stable targets.[39] The CIA's Counter Terror Teams in Vietnam (tellingly referred to as *Count* Terror Teams) relied upon what was called the Census/Grievance and Aspiration program, or Census/Grievance. While the census as a tool for counterinsurgency was supposed to provide the village head count, it also gave cover for collecting useful military intelligence. During this crucial period of the war (1963–67), military historians conclude that the Census/Grievance program "was not a critical contributor to the success or failure" of pacification. In the "lessons learned" portion of the U.S. military's own account of this lost war, there appears to be a missed opportunity for creating a better COIN doctrine. Count Terror Teams wished for better means for sorting the complexities of anthropological knowledge where the shifts in potential enemy movement and associations proved to be too large, and changed too rapidly, for traditional forms of census work to make sense of. This goal of providing the right combination of the knowledge of bodies, as well as of minds, is more recently being realized with greater efficiency and on a much larger scale than the old census technologies could provide. The "ubiquity of COIN thinking across the academic-military divide has contributed to a quantum leap of American literacy in irregular warfare."[40] This quantum leap has to do with using new tools, as the word "quantum" suggests: a move from minds and words to bodies and data.

Having jettisoned an old assumption where identities were more or less fixed, twenty-first-century counterinsurgency strategy uses digital counting and immediately updated databases rather than the slower print-based tools to process the complexities implicit in globally networked forms of war. The goal in the new COIN is not the deployment of prophy-lactically mediated contact with the host population-cum-enemy-cum-object of sexual desire. Rather, *terrain*-ing humanity finds a way to weap-onize communal complexity or, as in the increased use of robotic warfare on the order of drones, to target clusters and anomalous movements from high above, and kill without any human contact at all.

DATA AS PHYSICAL TRANSMISSION

The varying epistemological capacities available across alternating media forms—oral, print, and digital—are nowhere more provocatively engaged than in the work of Walter Ong. His observations remain extraordinarily compelling in the context of tracing media changes within military anthro-pology. By using Ong, we can better trace the move in war anthropology, this time not from physiology (Nazism) to culture (Mead), but from cul-ture back to bodies, and further, from bodies to physical matter at large. We are still focused on recasting humanity as terrain. In Ong's terms, this recasting is technology-dependent, and would eventuate in correspond-ing changes in the "spatial imagination."[41] We can apply this emphasis on spatially determined forms of knowledge as part of cultural rematerializa-tion for the purposes of war. For Ong, a physics of information marks the historical change from an ancient scholastic emphasis on rhetorical elo-cution, where the influence of words is primary over the accumulation of empirical data, to an Enlightenment model of the world where real things exist in scientifically accessible ways.[42] The arts-course scholasticism of the pre-Newtonian mind was founded on a disciplinary arrangement geared to process "audile" (sound-based) communication. The different media tech-nology giving rise to Western modernity was based on reading and writing, words seen both as things and as reference to things insofar as words are arranged on the page. By moving from sound to the physicality of print, writing became corporeal in a sense humanists and literary critics still re-fuse to see.[43]

Ong's work has lasting originality because he refuses this refusal. Ong forbids a simple dichotomy between, on the one hand, (Western) print media as responsible for producing modern subjectivity, while, on the other hand, (non-Western) oral communication committed to a primitively animated world of things. One of the most challenging and original aspects of this work is Ong's emphasis on the matter of Western writing, its corporeal transmission, in the most challengingly literal sense.[44] His work refutes the division between digitization and the most ancient types of communicative media—for example, microchips and Sumerian tokens, which were "clay figures (cones, spheres, disks, and cylinders) constituting numerically discrete units used to process data for reckoning purposes."[45] The effect of Ong's corporeal theory of media is not only to return modes of communication to their proper computational origins but also to move the question of communication into a problem of physical space. Here computation allows a kind of geologizing of the word. The Sumerian token, and the microchip, because they are both "three-dimensional," are closer in kinship than the "two-dimensional" surfaces of print. They are both "discretely numerable . . . and manually separable from one another."[46]

Further still, by expanding the historical focus from written communication within an abstract public-sphere ideal to a problem of materially existing "spheres" (literally, Ong's tokens took this shape), space itself "becomes more spacious." Again, what we may think of as *spaceless* endeavors (thought, culture, communication) becomes something that exists on the order of matter.[47] Further from Ong: "There is nothing more natural for a human being than to be artificial."[48] Thus for him, prophetically given the shortcomings of old COIN doctrine and the coming of the newer kind, the human being is a "biological information system."[49] Human relationships both require the use of media and can be construed as media. Ong describes social organization as an "evolving universe of quantities" and an "assemblage of flooding . . . swelling . . . electrons . . . in great masses."[50] This decidedly computational vocabulary, which culturalist versions of anthropology obfuscate, begins to intimate how life-worlds turn into terrain.

But Ong's emphasis on the Newtonian impulse to get at things as they are refutes the notion of a historical break between physically based forms of knowledge and the merely cultural kind. Instead of a notion of modernity as rupture with the past, Ong's quantified conceptualization of media is more akin to Latour's search for a modernity we never had: "We have never been modern" in the sense Habermas would have liked us to be. Ong's alternate modernity (and Newton's) emphasizes "not the human consciousness"—or again after Mead, "psychic unity"—"but the ranging of objects in space . . . a mathematical physics . . . , and the production of categories with a strong mathematical torque."[51] Such a notion of "the ranging of objects in space" has application not only to print technology but also to all media, whether it is the latest in quantum computing or the twenty-thousand-year-old animal paintings of the Lascaux caves. To put Ong's emphasis on "the quantified manner of conceptualization" back in play on the topic of war means the study of human relationships should not be cordoned off from a physics of information.[52] With the force of Ong's mathematical torque, the difference between war and peace becomes a cultural affair, but with the key provision that we do not read the term "culture" as a localized subjective abstraction.

Print-based communication was the preferred media for fighting small wars in the 1940s, and for reasons privileging a cultural over a mathematical rationale. The need for print to slow things down for war to catch up is key here. In contrast with how Ong would quantify all communication, we read in the *Field Manual* that "current periodicals and newspapers" are contrasted with "the 'underground' or 'grapevine' method of communication."[53] Print media is deemed the most efficient way to achieve tactical control of the insurgent's spoken words. "Transmitting information" splits here between rumors, which spread with "unbelievable rapidity among the natives," and the occupier's need to put events in a properly serialized and homogeneous order.[54] Similarly, T. E. Lawrence comments in "The Evolution of a Revolt": "The printing press is the greatest weapon in the armory of the modern commander."[55] Writing rules in opposition to insurgent numbers. Thus, Lawrence puts the weaponization of print in this way: "The Arab mind," he continues, must be approached according to three "tactical elements": "one algebraical, one biological, a third psychological" (7). According to the "algebraic factor," the so-called Arab mind "would take first account of the area he wished to conquer, which required idle calculations of how many square miles . . . and how the Turks would defend all that" (7). Over and above the comparatively idle mathematical thinking that Lawrence links to physical corpora, the Western way of war must also apply psychological methods: "We had to arrange their [our men's] minds in the order of battle, just as carefully and as formally as the officers arranged their bodies: and not only our own men's minds, though them first, but also the minds of the enemy, as far as we could reach them" (11). The division between mind and matter, and the essential role military technology would play in maintaining, as well as challenging, this fragile separation, is a problem of computational tactics.

NATIONAL CHARACTER STUDY AND WORLD WAR II

What print media delivers for Lawrence, then, is the same as what the 1940s *Field Manual* sought: the proper sorting and targeting of revolutionary mixing. But the algebraic factor presents a permanent threat. Revealingly for Lawrence, and important for tracing the historical development of U.S. COIN doctrine, "algebraic factors are decisive when victory rests with the insurgents" (22). In Lawrence's case, mathematics overcame a culturalist resistance to numbers whenever the Arab mind got the better of Western subjectivity. Despite the tendency of war planners to forget it, the *quantitative* features of insurgency were never far from the *qualitative* ones developed to subdue them. Again, this is not to say that the division between orality and written communication existed absolutely in the 1940s COIN theory. Rather, the computational grapevine and the psychologizing printing press are each evaluated in terms of delivering the right amount of usable data. The native problem of too many messages in too many directions creates confusion because it operates according to a media form unattuned to a print-based mode of subjectivity. For the counterinsurgency soldier in Lawrence's time, the success of print as a war tool lies in its ability to calibrate multiplicity into stable sets of informational pairings, or at least, such is the hope: self versus other, past versus present, Arab math versus Western intellect, materiality versus culture, and so on. Data left outside these oppositions will presage a different tale than the one anthropology tells in terms of morals and morale. The story of the insurgent is silent because there is no way for print to record it. But silent data become visible through a physics of information, apropos Ong. Here the human-versus-terrain divide oscillates between one and the other, putting a new twist on the culturalist version of anthropology offered by Mead. Human relationships become the object of the mathematician, the engineer, the geographer, and the physicist, no longer merely the source of national character, so called.

Given the interoperability between media, matter, and mind, there is dissatisfaction with the term "culture" in posthuman war. Twenty-first-century COIN doctrine finds the concept vague and insufficiently concrete for mapping the complexity of the network-centric battlefield. In the U.S. Army's groundbreaking 2006 *Counterinsurgency Field Manual* (FM 3–24), the words "culture" or "cultural" are mentioned 178 times in just 282 pages. But even with a surge in the use of this term within the new COIN doctrine, defense experts remain resistant to a purely anthropological approach to war without appropriate computational correction: "The cultural turn is empirically unviable," they write, "and politically naïve."[56] Echoing a consensus within mainstream academic ethnography against subjectivism, another anthropologist refers to military uses of terms like "ethnic" as "shrewdly essentialized," based on "outmoded anthropological models," and ultimately expressive of "the white man's burden reinvented for the age of identity politics."[57] The outmoding of cultural studies underlying these objections hearkens back to a divide between the study of national character and the search for more empirical ways to surmise the contours of insurgency.

To look backward toward an earlier mode of war anthropology once again, and just as a matter of emphasis, we should recall that no less than 50 percent of professional anthropologists in 1943 worked full-time in the war effort. Another 25 percent did so part-time.[58] In the months after the Japanese surrender, the U.S. Senate's Committee on Military Affairs heard testimony for increasing the number of anthropologists involved in future war efforts. This expansion of the discipline was expressed not just in terms of increasing the numbers of practitioners but as a quantitative problem on its own terms: war planners wanted more data for understanding "whole great cultures." The key point is that wholeness was then defined as too vast for the intelligence at hand: "There was not enough of them [anthropologists] to provide more than a fraction of the information needed."[59] The interest in how fractions might to add up to wholes for purposes of fighting wars is worth underlining. This is because the persistent problem of how war scales up lasts into the future of war anthropology in its postwar years. Now war becomes more ubiquitous then ever and has no clear period of beginning or ending. At the same time, fractions are addressed by the computational push against culture in its very name. In turn, the goal of joining political force to the physics of information seems reachable at long last.

But the war planner's desire to see human relationships in this materialist vein can already be seen in anthropological work sponsored by the Office of Strategic Services (hereafter, OSS) during World War II. The OSS, which is the institutional forerunner of the U.S. Central Intelligence Agency, employed twenty-four anthropologists in 1943.[60] In that same year, the agency released its "Preliminary Report on Japanese Anthropology." Race, not culture, was the key word here, and it was used in a biological form. As David Price points out, one Harvard anthropologist recommended the study of Japanese prisoners and interned Japanese U.S. citizens as "specimens of constitutional study [helping to] yield useful information regarding the weak spots of the Japanese physique."[61] The notion of constitution here is a corporeal one, not yet geological in the sense of human terrain, but nonetheless material for its emphasis on physique. Inner-ear structure, taste buds, laryngeal musculature, and other morphological features of the Japanese body were examined in the interest of "introducing some disease among enemy troops that might catch them by surprise."[62] According to one report, measurement of the intestinal track revealed Japanese men, more than women, "were liable to inflammation or ulceration of the peptic type."[63] This characteristic of the Japanese immune system could be advantageous in a biological attack, such as the application of anthrax or other agents on the battlefield.

As Price points out, Japanese anthropology in 1943 was not exclusively limited to the more banal sounding study of character. It "included the contemplation of biological warfare programs using anthrax and other weapons of mass destruction on Japanese civilian and military populations."[64] As the OSS suggested, "The possibility of spreading infections of various kinds to attack the respiratory tract, a known weak spot in the Japanese body, would be the most effective agent."[65] As with the biological treatment of the enemy as racially vulnerable, Western conceptions of physical difference run parallel to a temporal narrative about the progress of civilization. Space and time are different for the Japanese, and tellingly, this is because of their misuse of "our technologies" as per Mead's machines and Lawrence's algebraic factors.

In 1945, the Sydney *Daily Telegraph* referred to war within the Empire as "confronting 2000 years of backwardness against a mind which, below its surface understanding of the technical knowledge our civilization has produced, is as barbaric as the savage who fights with a club and believes thunder is the voice of God."[66] Note that in the first citation, the racial other is portrayed as a contamination from which the opposing race must be made immune. More enigmatically, with the second quote in mind, we see how the direct use of one form of contamination ("introducing some disease") to get rid of a second one (a contamination between East and West) belies the presumption of how racial divisions are made. They are made with "technologies" of war. If race is biologically constructed, as the first quote implies, it can be objectified, cordoned off, and annihilated by the application of biological force. If "technology" is doing identity work, as in the second quote, and the work is going on in ways spanning East and West, then the crossing between races makes the enemy hard to define. Everything is coordinated by the distinction between their clubs and thunder and our civilized fighting tools. Not surprisingly, the other-ing of Japanese identity in the *Daily Telegraph* depends on the opposition between East and West as a technological matter. The newspaper itself is the technology at work in helping to identify how the enemy might wield its less sophisticated club. But there is a more significant point for understanding the *pre*-history of *post*-human war, which is twofold: the friend/foe opposition has always been a technical one, and technology strains the opposition between culture and the human body.

According to the "cult of the sword" described in *The Chrysanthemum and the Sword* by Ruth Benedict, who headed the Office of War Information in 1944, "the Japanese were the most alien enemy the United States had ever fought in an all-out struggle because it did not belong to the Western cultural tradition" (1–2). Benedict wanted to "look at the way they conducted war and see it not for the moment as a military problem but as a cultural problem" (5). In her emphasis on the culture-war connection, Benedict is especially resistant to numbers and empiricism, and we see this resistance, as is consistent in the period, put in terms of technology's dismissal. Concurrent with Mead, Benedict writes that the cultural anthropologist must distance herself from the "stock technique of sociologists who amass . . . census materials, great numbers of answers to questionnaires, and scientifically selected samples of population, which have been highly perfected in the United States" (17). The scientific challenge of amassing the features of human relations as mere census counting is to be replaced not by expanding data but constricting its relevance as old hat. For Benedict, "qualitative study" displaces stock techniques for amassing mere numbers (18): materials are supplanted by subjective forms of value; science by cultural work.

But consider, too, the case of Geoffrey Gorer, Benedict's predecessor at the Office of War Information. Gorer wrote the first national character study in 1942, which also focused on the Japanese soldier. He did so, unlike Benedict—or in a way less concerned with opposing cultural study with quantitative science—by focusing on identity and biology together. In this way, his work is consistent with the juxtaposition of citations from the OSS and the *Daily Telegraph* above. For Gorer, "severe and continuous punishments" associated with the "toilet training" of Japanese men is "the most important single influence of the adult Japanese character."[67] This emphasis on training the body, while not the same as the state racism of the Nazi regime, nevertheless shows the ease with which the term character could be regarded as having physical attributes. According to Weston La Barre, writing in 1945, "the critical trauma . . . at the anal level of development" was the source of the compulsive personality of the Japanese."[68]

These examples are interesting not simply for their tendency toward psycho-determinism but also because they mix biology and psychology together for the purposes of war. The "conformity to rule," "fanaticism," "sadomasochistic behavior,"[69] and so on, helped La Barre explain the unthinkable enmity of the Japanese soldier, who was violent to the extent that his subjectivity was alleged to be extinguished. The "character structure of the Orient" elicited a kind of war to be fought in an unthinkable way, unthinkable because the enemy did not adhere to anthropology's preferred mind versus body distinction. Japanese soldiers did not have the ability to transcend their prehistorical biological chains. Therefore, paradoxically, they could not uphold the values of self-preservation. Because they were without morality writ as individual worth, the Japanese could bear extermination without the exterminators bearing moral cost. To cite Price once again, the "Strategic Bombing Survey used psychologists, anthropologists, and other social scientists to analyze the impact of Allied bombings on enemy military and civilian populations."[70] But the culturalist approach to war shared a flaw with the biological reductionism it sought to oppose. In both instances, a presumption of unity across differences was overlooked to produce a disciplinary division corresponding to a geopolitical one: culture and subjectivity were distinguished from science and numbers, mirroring the opposition between Allied and Axis military forces.

In an era where national character study became the dominant anthropological model against the biological racism of the Axis powers, remarkably, U.S. national security policy held fast to a race-specific rationale for killing both Japanese soldiers and civilians. More remarkable still is the more general set of points: despite the tactical investment by U.S. war planners in national character study, culture did not offer an effective escape from biology in World War II. Subjectivist claims were remained intertwined with physical ones. For all the racial forays into studies of the body by some U.S. anthropologists after the Japanese attack on Pearl Harbor, the OSS assured its researchers of the importance of preserving Western cultural values. Whatever race-based weapons might eventually be needed to win, all moral and ethical implications would be considered.

Exemplifying how quantitative versus qualitative war anthropology addressed the problem of the enemy at home, the evacuation of 110,000 Japanese Americans starting February 19, 1942—two-thirds were U.S. citizens—was in the first instance an exercise in the power of census-taking. Here, too, biological objectification was underwritten by a cultural rationale, as establishment educators praised the quality of camp schools, and as social workers attested to the civilizing benefits of race-based relocation.[71] In the 1920s, James D. Phelan, Democratic senator, and chief financial backer of the Oriental Exclusion League, could use white supremacy to save the state of California from a form of Japanese "mongrelization and degeneracy," lumped in with the rise of Bolshevism. The American Legion's resolution calling for total imprisonment of all persons of Japanese ancestry specifically used the idea of "race relations," even though independent community analysis reports indicated a "morass of conflicting attitudes" toward American loyalty among the U.S.-born Nisei generations.[72] In this sleight of hand between biology and culture, "American citizen" becomes "Japanese foreigner," and through the application of census-based political mathematics, the definition of the enemy is both widened and contained.

The point of contrasting the rules of race with the morass of attitudinal data is not to suggest the traditional next step, which would be to say, in the manner of Benedict and Mead, culture trumps numbers, subjectivity precludes technology, and a fully dematerialized approach to human relationships is the only knowledge capable for achieving universal values. The "social self-consciousness" sought by Mead as early as 1942 was at best an idealization of the qualitative part of humanity celebrated by her colleagues at expense of the quantitative kind (3, 161).[73] But we should recall that the title of Mead's book—*And Keep Your Powder Dry*—is attributed to the seventeenth-century Puritan revolutionary warrior and regicide, Oliver Cromwell, who said: "Trust in God, and keep your powder dry" (1). National character, like anthropology itself, has never been far away from war. War is inextricable from the comparative study of human relationships even in its most culturalist guises. But this is finally a less dramatic claim than what takes place after we admit culture's proximity with bodies in the history of war anthropology. Recovering its physical underpinnings, so-called character allows us to compare and contrast old and new ways of doing cultural-as-tactical work. Mead's national character studies helped wage a war not just between national ideologies: the biologically perverse ones of the Axis powers versus the culturally liberal ones of the Allies. She was simultaneously addressing a series of related divisions: subjectivity over objectivity, the experiential over the physical, narrative over computation, stories over technology, humanity over matter, and so on. The winning of the first term over the second in each of these pairings would determine what kind of knowledge postwar anthropology would become for succeeding generations. But something else was lost in the winning.

The anthropological desire to dismiss human community as blood in favor of community as upbringing brought with it a turn toward qualitative forms of study (cultural, moral, and character-based) and a turn away from quantitative tasks (computational, numerical, and physical), as we have already seen. The antiracist intent of "finding consistencies and regularities in character formation'" (21) is clear. But in the same way that Mead's involvement with war put new pressures on humanistic knowledge, war anthropology after World War II once again changes the definition of humanity. In posthuman war human relationships are rematerialized, and the number technologies that cultural anthropology used to forbid takes on a central role. For Mead, anthropology was born not only out of war but also out of the expansion of war, and from there, out of disciplinary-qua-military change: "In moments of crisis, in order to keep one's powder dry we must be accurately informed. What are our strengths and weaknesses?

If we have a weak flank let's not leave it uncovered. If we have a shortage in material needed for ship building, then let's not put our faith in ships" (60). For Mead, a crisis of faith in human relationships corresponded to a crisis of faith in how we produce knowledge about them. But to maintain Allied culturalism outflanked physically based forms of Axis political violence would be misguided. Culture and biology worked hand-in-glove across the Allied versus Axis divide. What gets outflanked is less about culture than the culturalist's presumption of working outside a system of war neither strictly human nor exclusively national in nature.

COUNTERINSURGENCY THEORY AND THE VIETNAM WAR

The crisis of culturalist war anthropology in the 1940s provides the right context to return once more to the U.S. war in Vietnam. In 1962, Defense Secretary Robert McNamara offers a lament that will lead him back to an old preoccupation with numbers: "The Communists have an enormous lead in Revolutionary theory."[74] Those revolutionary leads meant that U.S. Cold War adversaries were perceived to be more adept than its own military strategists at mobilizing "what Mr. Khrushchev calls wars of national liberation or popular revolts, but which we know as insurrections, and covert arms aggression."[75] The key word in McNamara's lament is "popular," because it continues to mark a long tradition in war history where a problem of numbers is dismissed by evoking a peace versus war distinction along Kantian public-sphere lines: liberation is unthinkable when citizens use political force within a sovereign nation. Further, from McNamara: "The occupation army of an imperial power is always outnumbered by the indigenous population, and [this numerical imbalance] is offset by the technological advantage of the occupier."[76] Here one kind of numbers problem, the insurgent outnumbering the occupying army, is addressed by a technical solution. One might say, McNamara attempts out outnumber the problem of outnumbering itself. His lament is not so much against having too many insurgents to deal with but in not being able to get the data on his side.

Already by the 1960s, and foreshadowing a need to think again about anthropology as sympathetic with far-reaching forms of data analysis, Defense Department contractors expressed frustration with early attempts to use soft science for military goals: "When you can express it [knowledge about foreign communities] in numbers, you know something about it; but when you cannot express it in numbers, your knowledge is of the meager and unsatisfactory kind."[77] Washington has since demanded better technologies for counting populations in war and has long decried the imperfection of data.[78] But the demand comes with familiar tensions: "Social scientists strove mightily to add quantitative statistical aspects to their work, [but the] Viet Cong prisoner's claim to have joined Communist forces seemed very personal."[79] Thus the attraction to numbers as well as the lament over them in the McNamara years has proved tragically frustrating. It inspired a slogan within military policy-making in later years to "replace 'mass' with 'information.'"[80] McNamara's inclination to turn everything in war into data created the complication of having too many data points. This only further embedded the distinction in COIN doctrine between soft power and what could be counted as the harder stuff. The mass was not yet rendered compatible with knowing something. As McNamara notes, "I had never visited Indochina, nor did I understand or appreciate its history, language, culture, or values. When it came to Vietnam, we found ourselves on terra incognita."[81] For culture to have been more fully integrated with his desire for war to better utilize, and effectively counter, revolutionary numbers, COIN doctrine would have to wait a generation for the HTS program.

A 1964 document, *Applied Analysis of Unconventional Warfare*, serves as an early attempt at the weaponization of culture along the eventual lines of human *terrain*-ing. But it is also an example of the how disciplinary frustrations continued to play out after World War II in the science-versus-humanities divide.[82] The hope in this document on "population centric warfare" is "for scientific analysis to be useful in suitable forms of intervention" (1). And characteristic of later U.S. COIN doctrine, "intervention" is defined as "gaining control of the enemy's civilian population" (1). However, the 1964 document immediately turns to the problem of "data handling," setting computation up as a problem as yet to be fully resolved: "One of the striking experiences of this [quantitative] research has been a certain shock upon realizing how deficient is the range of data available for modern societies with literate traditions" (5). The strike referred to here—as in a "striking" experience—is the shock of numerical scale still exceeding the intelligence of war anthropology. The fluidity of movement between civilian and insurgent activities, and the speed with which these relationships change, proved "surprisingly complex" (5). Moreover, the word literacy is juxtaposed with data, putting the question of media at the center of Vietnam-era COIN doctrine, and echoing Lawrence's frustration with the Arab's algebraic intelligence. The document thus continues, "The 5 by 8-inch end-punch descriptor cards, because of their continuous expandability and their adaptability to multiple indexing schemes, have proved particularly useful in systematizing accumulated data" (5). As the words expandability and multiple imply, computational technology promised an ability to scale up information so as resolve the striking complexity of insurgency movements and reveal more accurate and timely ways to launch a response.

Nevertheless, as a related document reflecting back a decade later reveals, "The basic data were simply not good enough for . . . setting up a computerized data system for the operators to use."[83] The fundamental importance of "a complete census" is emphasized in the 1960s, but it too runs into problems over the reliability of information processing on a numerically vast scale. Indeed, one of the frustrating parts of combating communists in Vietnam was that they fought the census and refused to be identified by any of the categories listed there: "Begin by taking a new census," one "Strategic Hamlet" report reads, though it "often prompts some pro-communist groups to flee the hamlet."[84] So, too, on the so-called Republican side, "social mass movements such as the Republican Youth, and the Vietnamese Women's Solidarity Movement, suffer from popular indifference and lack of identification."[85] The "enthusiasm" expressed in the Vietnam era for adjoining such disparate fields as "anthropology and mathematics and disrupting normal disciplinary lines"[86] would have to wait for a technological update of U.S. COIN doctrine still to come.

This update is evident within the U.S. Deputy Secretary of Defense office's initiation of Project Maven in 2017, which seeks the "establishment of an Algorithmic Warfare Cross-function team."[87] The Project Maven memo reads: "Although we have taken tentative steps to explore the potential of artificial intelligence, big data, and deep learning, I [the deputy secretary of defense] remain convinced that we need to do much, much more, and move much faster, across DOD [Department of Defense] to take advantage of recent and future advances in these critical areas" (n.p.). The emphasis in "much, much more" should be read as an indication of the desire to scale up war through expanded computational means. In response, the University of Maryland's Institute for Advanced Computer Studies has devised a SOMA (Stochastic Opponent Modeling Agent) portal also using computer science to enhance human observation of social patterns. Using "accurate behavioral models and forecasting algorithms," SOMA offers "a virtual roundtable so terrorist experts can gather around and form a rich community that transcends artificial boundaries."[88] As SOMA documents make clear, the boundaries diverge and are many. There are the boundaries dividing academic disciplines like science and humanities, and the boundaries dividing academic knowledge from military force; at an even more important level, there are the boundaries subject to a necessary computational re-rendering of merely human cognition about the tempo and geographical relations of the planetary war.

The SOMA documents propose to transcend artificial boundaries of these kinds by utilizing technology in new ways, merging "computer science and social science." The consequence of this is to merge representation and empirically observable human activity, fusing virtual and accurate behavior. Added to this change accessing spatial relations, there is a corresponding acceleration of time: SOMA software is predictive in the sense that it allows for preemptive military strikes. The technology is geared for what might be called contentless war, as in war where the identity of the enemy is open and all-inclusive. This is indicated by the first word in the titular phrase: "Stochastic Opponent Modeling." Stochastic means "randomly determined; having a random probability distribution or pattern that may be analyzed statistically."[89] It is a word taken from the Greeks, *stokhastikos,* meaning "to aim at, to guess." To aim in this way is to make a future reality present by assembling data in a new way, to pattern out intelligible knowledge from what might superficially look like terra incognito (recalling McNamara's term for revolutionary numbers in Vietnam). When "computer science and social science combine," data scattered too widely for human observation to capture, as per the frustrations of the census takers in the jungle, now become available to war. Moreover, war is extended to put all undesirable groups on equal footing with the enemy. In that sense, "SOMA introduces a paradigm for reasoning about any group that is a terror group, a social organization, a militia, or an economic organization."[90] Any form of "political behavior, rebellion or protest" can be made legible by SOMA in order to monitor, control, and respond in real time.[91] The only requirement for being aimed at in a stochastic way is being organized, even if such organization is not something you intend or are aware of in advance. If the data you generate can be legibly tallied in a future-oriented direction, pointing toward some pattern of association deemed dubious now or at a later date, you are already targeted within the SOMA portal.

U.S. Special Forces officer and war theorist Patrick Duggan embraces the "dawn of machine intelligence" in this way: "Success in warfare is not derived by labeling its assembled parts. It is achieved by correctly assembling technology and new capabilities into trusted fields of practice."[92] The expanded battlefield itself is the new field of technical practice. But the further-reaching point of this statement has to do with the assembling (and reassembling) of parts at increasingly expansive levels of scale, and rendering what once seemed mere fragments of information into relevant patterns for military goals. The opposition between friend and foe are blurred when civilian associations are read concurrently as a security threat, as SOMAs promise to target social organizations and terrorist groups in the same way. The reassembly of human relationships along more expansive lines such as these is where war anthropology adjoins computation, pace Benedict and Mead. On the Special Forces embrace of machine intelligence, data literacy is the most special force of all.

The oscillations between culture and bodies in classic COIN doctrine, running from the depictions of Native Americans as foreigners, through the race-based imprisonment of Japanese American citizens and into the calculation-defying actions of the Vietcong, gains new intensity with the more recently invented "cyber-enabled Counterinsurgency" (CE-COIN). Duggan, in "Man, Computer, and Special Warfare," notes that CE-COIN utilizes "Cyber, Computer, Communication Hygiene, and Non-Standard Communication" (5), opening human relations to technology in an unprecedented way. As the association between communication and hygiene suggests, media blurs the distinction between cultural and material relations. CE-COIN emphasizes not only the expected areas of computer technology but also new forms of military cooperation between bodies, media, and mind. CE-COIN moves seamlessly between "Humans and the Physical World," and does so, as Duggan suggests, through "experimentation." We the human being—or more accurately, we the human being a computational entity—are the stuff of the experiment. The exchange between biotic and non-biotic systems is made possible here, as in the language of stochastic analysis, by using "virtual systems and open networks" (5). And in the same manner of SOMA technology, space and time become malleable in ways highly useful for targeting the insurgent in the most expansive sense of that term. Duggan continues, this "wide array of communication, command and control, [as well as] mapping, meshes devices into network partners. CE-COIN moves from the presence of a known enemy to its possible or likely 'tele-presence'" (5).[93]

To the extent that network-centric combat at its most extreme remixes relations between humanity and machines, the experimental phase of war anthropology decivilizes the population as is generally consistent with posthuman war. But Project Maven, SOMA, and CE-COIN are particularly significant because of their capacity to subsume all phenomena within a battlefield of unlimited reach. Each of these programs rely upon "computer vision algorithms for object classification, detection, classification, and alerts."[94] In this mode of classification, the term "object" is co-equal with target. As such, "object" commonly designates all persons and all things. Any material or immaterial arrangement can be sussed out of an otherwise incomprehensibly large set of data points requiring computation to be known. The object to this extent is precisely phantasmatic in the sense of CE-COIN's new emphasis on tele-presence, writ here as unseen relations within seen relations, or "virtual constants in intra group conflict."[95] The term virtual here designates a space–time dimension of war not available to culturalist models of anthropology. But virtual is nonetheless real. It is simply the kind of reality escaping human forms of recognition as so much *more* than we see. It is a reality exceeding the realism projected by the inferior media composed of human retinas and light. The ability to move across and within discrete categories of organization serves to undo, and then rearrange, "objects as alerts." In this sense, CE-COIN means the end of objectivity, or at least the end of an idea of objects as neutral or immutably there. Thus, posthuman COIN doctrine transcends the difference between kinds of knowledge while recombining the difference between people and things. Taken on the largest possible scale, all human behavior is precriminal behavior. In the eyes of the new war machines; every object is also a threat.

The term "maven" in Project Maven, which is of Yiddish derivation and fully entered the English lexicon only in the 1960s, is revealing. As it suggests, the twenty-first century's experts in human relationships are not altogether human. The notion of all objects as alerts ushers forth new targets from a previously inaccessible universe of people and matter and makes no distinction, or no permanent one, between friend and foe, person and thing, subject and object, mind and machine. I say no permanent distinction because virtual reality as an application of war presents the capacity to shift seamlessly between human and nonhuman terrain. The shift itself is therefore best described as an oscillation, because if virtual war technology is working at full capacity there is no difference between what is there in a telepresent way and what you regard as physically present to the human eye. Indeed, what is at stake in posthuman war, as it moves with the least possible friction between visible and invisible object-alerts, is the ultimate achievement of eliminating the split between representation and the reality. Again, this requires a special kind of communication, not communication of the suspicious oral story, nor even the more trustworthy printed letter, but numerical communication instead. War moves from a movement between culture and biology at this point to a movement between computation and matter. Insofar as the virtualization of the battlefield in CE-COIN subsumes rather than simply displaces the earlier applications of COIN, technology and the physical world move toward a radically "nonstandard" point of convergence. War invades civil society as an informatics of military violence. But war also invades the means of communication, no matter what the communicator means. In short, *anything* perceivable as data can be weaponized. This is a special moment in the history of war because *everything* can be perceived in such a way.

The *2015–2016 Assessment of the Army Research Laboratory* report offers a series of statements reflecting the convergence of the physical world and war—through numbers—in ways consistent with Project Maven.[96] The document is a product of a revealing mix of disciplines, with subtitles like: "A Report of the National Academies of Sciences, Engineering, and Medicine." Like the term "communication hygiene" in CE-COIN, the adjoining of medicine with engineering raises the prospect of human

relations being manufactured within an unusual form of the military industrial complex. A keyword in the *Assessment* document is "bio-inspired
materials" (xiii). The material comport of military bio-inspiration on this
order represents an epistemic frontier where medicine and engineering
converge. Under generically capacious chapter headings like "Energy and
Power Materials," "Engineered Photonics," "Disruptive Energetics and
Propulsion Technologies," "Multi-agency," "Electronic War," and simply,
"Data-Intensive Sciences" (xiii–xiv), culture becomes bodies; bodies become war machines.

This is exemplified by the first ribocomputing technologies, which use
synthetic RNA (ribonucleic acid) to sort through "multiple biosignals."[97]
The effect of this sorting, based itself on a new kind of animate material,
can in turn make "logical decisions to control protein production with
high precision," and will have a dramatic impact on everything from medical treatment to the production of biofuels. These products come with molecular level "triggers," a term that belies the separation between ribocomputing and military violence. As the phrase Living Foundries suggests, life
is reoriented as a product of war manufacturing. Accordingly, discoveries
most valued by the Army's *Assessment* report war research reach beyond
cultural realms defining war anthropology in the World War II era. As it is
now being "scoped," the *Assessment* remarks, "the area of humans in multi-
agent systems is very broad. It includes interaction between humans and
technology" (9). The most bio-inspirationally informed military research
will thus "focus on mission-relevant problems and contexts and draw on
measures from multiple domains (e.g. bio-mechanics, cognitive theory,
and neurosciences)" (164).

CE-COIN research reconceptualizes the human body as an engineering problem, and more specifically, renders subjectivity into technology
not just with computational skills but as a computational entity in its own
right. The "statistical power of the EEG signal in the brain" enables army
researchers to "provide feed-forward control signals for exoskeletons,"
and other "highly mobile sensing systems that could be useful in the field"
(165). The word "system" is perhaps now one of the most important concepts in military research. Its ability to move in coherent and empirically
calculable ways between biology and physics, technology and communication, people and machines, "atmosphere" (7) and public sphere, makes

"system" an exceptionally useful concept.[98] It works the same way the more familiar military buzzword used in the RMA—"network"—works. But does "system" refer to the mode of representation of any given physical reality? Or does it refer to physical reality itself? The answer, surprisingly, is both.

As applied in posthuman war, the concepts of system and network provide effective ways to move between, and reintegrate, the realms of mind and matter. This happens on the order of what the *Assessment* calls "device physics"—for example, infrared vision, or any number of interfaces between the human and machinic entities (3). As with Ong's corporeal theory of media, "system intelligence" purports to be capable of "extracting relationships from diverse texts for constructing knowledge networks, and using agent-based semantic analysis for information retrieval" (5). The significance of the word "text" used by the army *Assessment* document is important. The agent doing the reading of text in this case is also a textual entity, only the text at hand is not only letters but also, as important, numbers. Information here works at a decidedly physical level, where there is no separation between sensing, knowing, and targeting. For example, "cross-modal face recognition uses new sensor algorithms . . . fusing data from multiple [i.e., human and computer] sources" (6). The words "system" and "network" work together—as in extracting as well as producing—a given reality in virtual and empirically based ways. Networked war systems proceed from a point of convergence between subjective and objective ways of achieving "object discrimination" (6). But those objects are only as real as they are rendered visible—or to use the right lexicon, scoped—by way of intelligent machines. More significantly, networked war systems calculate a single and totalizing battlefield from points of reference too large for human beings to know.

QUANTUM SYSTEMS AND ASYMMETRICAL WAR

Nowhere is the U.S. Army's socio-technical front line more likely to be pushed ahead than in the *Assessment*'s interest in "quantum methods" (6). The goal here enhances access to networked relationships and increases data speed, since quantum dynamics are consistent across molecular structures, and quantum computing is millions of times faster than digital technology. "Beyond establishing modeling and simulation capabilities," the *Assessment* declares, "advanced computing architectures in quantum computing allow fast cross-architecture execution." The crossing in mind here refers to "parallel processing environments for large scale heterogeneous parallel systems" (7). How large are these heterogeneous systems? All the way large: minds, bodies, geology, atmospheric flows, any event perceivable as information exchange becomes a matter of quantum networking. This could be as large as the physical universe or as small as a single atom. Any event related to any other can be systematized if the computational power is sufficient. Quantized reality in the *Assessment* document means understanding all physical phenomena in an empirical, logical, and militarily coherent way. This not only moves traditional war anthropology away from the vagaries of culturalism, it also moves human relationships into larger forms of material association. Identities (think genes and memes), stars (think nuclear fusion), atoms (think the movement of protons and neutrons), all become what quantum researcher David Deutsch calls "entities." All aspects of the physical universe may be quantized as "channels of information flow."[99] If an entity obeys the laws of physics, as Deutsch says in "What Is Computation," "it is computationally friendly."[100] Moreover, if nature does anything, "nature computes" (551). "A computation," Deutsch continues, "is a physical process in which physical objects like computers, or slide rules, or brains, are used to discover, or demonstrate, or harness properties of abstract objects" (557).

This statement about "computational universality" (558) is not as scientionally as it sounds, unless you are willing to allow for a productive relation between fiction and science, as anybody interested in transforming the world—war researchers and war resisters alike—surely must be. This is because there is a unique analytical language available to us coming from the quantum physics corners of posthuman war research. *The Department of Defense Dictionary of Military and Associated Terms,* for example, defines the "node" on the order of Deutsch, both in terms of information exchange and as a "physical switching location."[101] In line with the other ubiquitous keyword from CE-COIN and SOMA documents, this knowledge-generating relationship is also a system. For the *Defense Dictionary,* appropriately and consistently in terms of the lessons of the quantum world, "system is a functionally, physically, and/or behaviorally related group of regularly interacting interdependent elements; that group of elements forming a unified whole" (183). Note that function over form is what designates the power of system, or better, system makes new groups out of any number of functions, even if those functions are not self-evident. System finds a way to make living and nonliving elements cohere, at least provisionally, no matter how apparently disparate, as long as they exist in some empirically identifiable state. Finding ways to adjust the scale with better (because more capacious) measurements across quantized relations, and thereby, directing complexity into knowable pattern formation, is where the relationship between tools and productivity comes in. It does not come in by way of some absolute notion of difference. Against precisely such a notion, the opposite of identity is not otherness, but numbers themselves.

In this sense, system comes under the U.S. military's *Defense Dictionary* as having a special information related capability. It is "a tool or technique employed within a dimension of the information environment that can be used to create effects and operationally desirable conditions" (125). Like the operating system I am using to type these words on a screen, the creation of information does not negate its empirical reality, though it does negate naïve empiricism. Thus the computational turn in war away from culturalism is not an example of technological determinism. To make this charge is to miss the cooperative relationship between technology and empirical reality, and risk falling into idealistic absolutes.[102] The elements of the physical universe are really there, and so is the productive technical capacity needed in order to know them. We just have to assemble and reassemble—that is systematize—infinite complexity in order to produce new knowledge. This is a high stakes concern insofar as new knowledge is now more than ever a matter of life and death. Along these lines, Clifford Siskin makes a key pronouncement about the generative capacity of systems. Siskin says that it is both a "way of knowing what is really there" and a "mode of production."[103] System designates the deliberate use of categories to reveal common threads between relatively autonomous domains. These domains, to stay with the *Defense Dictionary*'s vocabulary, would not be considered parallel if you were unable to bridge the gaps between information, people, and things. But computation is not only the tool used in CE-COIN to bridge those gaps, it is a view of reality itself as a form of computational exchange.

The term "parallel systems" is thus defined in the *Defense Dictionary* as
the ultimate horizon of military research. The focus here is less on identity
than what this text calls "positive identification." A highly relational and
fluid, but nonetheless real, form of "identification is derived from observation and analysis of target characteristics including visual recognition, electronic support systems, or other physics-based identification techniques"
(204). You cannot maintain traditional metaphysical categories like subject versus object, representation versus reality, or mind versus matter, seen
in classic versions of counterinsurgency theory, if you take seriously the
term "physics-based identification techniques." Identity intelligence makes
no formal distinction between individuals and networks. Both are composed of "identity attributes and are available for mathematical manipulation as objects belonging to a class" (121). The generic nature of identity
as understood here is not fixed in the same way the civilian's identity is not
fixed in relation to the insurgent's. The analysis of parallel systems reveals
differences within and between categories of identity, so as to change the
classification structure from within and then move infinitely outward. The
nonbinary nature of this kind of identity intelligence is precisely consistent
with the nonbinary nature of quantum computing.

Given the emphasis on computation as a way to make categories out
of scaled-up data arrangements, McNamara's frustration over finding the
right means in Vietnam is significant. His search for the right form of computation is a harbinger of quantitative (or better, quantum-level) counterinsurgency theory to come. The tools for sorting populations (masses) into
friend-or-foe distinctions (categories), and thereby moving between populations to individuals (targets), are changing in posthuman war. Tools are
also changing the value of the entities being categorized (human = nonhuman beings). The French Army fighting their insurgency in Algeria had in
excess of three hundred thousand men equipped with the most modern
military kit available at the time and fought—unsuccessfully—against an
insurgency of thirty thousand equipped with only light weapons. As Roger
Trinquier wrote in *Modern Warfare* about the French-Algerian war, and
with an eye on the U.S. invasion of Vietnam: "The manipulation of the
population was presumed to be the deciding factor."[104] Like McNamara,
Trinquier laments, while "we all know that it is not at all necessary to have
the sympathy of the people in order to rule them . . . the right organization

can turn the trick" (5). This lesson is offered to American war generals in Southeast Asia even alongside an emphasis on "the role of *social* services" for military uses (50, emphasis in original). But even with the term "social," Trinquier emphasizes spatial, rather than cultural, aspects of this work. He describes human relations as a "logistical problem" addressed by population "gridding" (67, 72). Here we see the origins of the Vietnam's strategic hamlet idea, in which "inhabitants of the . . . villages . . . receive a *census card,* a copy of which will be sent to the command post of the sector and district" (74, emphasis in original). Here the management of population intimates a numbers problem merging social relations and the movement of bodies, even as social services are deployed for military goals. The further-reaching point in the gridding of human relations is to note the primacy of numerical information. Bodies are information, and this kind of information outweighs the importance of information generated merely by minds. This is where traditional COIN theory moves from weaponizing culture to the manipulation of data in physical and virtual reality at the same time.

A National Defense Research Institute report from 2007, *Byting Back: Regaining Information Superiority against 21st-Century Insurgents,* thus calls for the Pentagon to create its own "registry census," as well as expand its monitoring of cell phones, and other electronic signals. The first two words in the document's title indicate innovative attention to digital media, where digital information equals physical force. The full title also indicates reworking old ground by updated means.[105] Such an emphasis on information superiority presents a new way of computing insurgency, going through the human body, rematerializing what used to be presumed as cultural and ushering in a change of status for the human being as such. Counterinsurgency theory's current practitioners are engineers, mathematicians, and computer scientists, as well as social science students. According to one engineering journal, the Pentagon's Defense Modeling and Simulation office is producing "an astoundingly sophisticated amalgamation of more than 100 models and theories" with the goal, as Roberto J. González notes in *American Counterinsurgency,* of predicting how certain populations might react to a "gun pointed in the face, or a piece of chocolate offered by a soldier."[106] The "performance moderator functions" at the core of such "simulations" are "physical stressors such as ambient temperature, hunger, and drug use; resources such as time, money, and skills; as well as personality dispositions such as response to time pressure, workload, and anxiety" (cited on p. 81).

The essentially spatial turn—or better, spatiotemporal turn—is inherent in how CE-COIN displaces traditional war anthropology's emphasis on culture. As González astutely observes, McNamara refers to the U.S. occupation of Vietnam as a "social scientists' war." But next to the radical shift in "the war on terror" (ii–iii), cultural studies remains overly culturalist. By contrast, Lieutenant Colonel Fred Renzi makes the case for Human Terrain System by referring to "ethnographic intelligence" as "terra incognita . . . the *terra* in this case being human terrain" (quoted on p. 38). The extended reach of CE-COIN is provided by the new intensities of counting—call it Census 2.0—which open access to human relations at a level of scale and precision unlike any time before. This is the sense in which Purdue University's Synthetic Environment for Analysis and Simulation works, "gobbling up breaking news, census data, economic indicators, and climactic events in the real world" (cited p. 83). The adjoining of the synthetic and the real, as an expanded form of census-taking, is characteristic of war anthropology turning from merely cultural issues toward larger-scale environmental orders.

WHITE AFGHANS

In posthuman war, human difference is no longer reducible to identity or cultural or its mooring within cultural groups. Rather, these anthropological mainstays are superficially retained as nimble and expansive kinds of war matériel. One veteran of the War on Terror writes: "I began to realize how Marine Corps sayings and popular phrases were a cornerstone of Marine self-image. Our cultural norm of expressing emotions and experience through socially understood sayings is reminiscent of the use of poetry by Bedouin women."[107] To "Marinize" the battlefield thus includes "the everyday, gritty reality of implementing a culture policy that requires . . . cross-cultural understanding, and even 'tea-drinking' into this ideal expeditionary Spartan warrior culture."[108] The move from human terrain as a geographically measurable field of permeable objects is not fully developed in these passages. But the words "gritty" and "reality" do indicate

an accounting for human relationships where everybody is, if not interchangeably, then at least flexibly open to unlikely associations. Specific cultural traditions can be encountered and more and less effectively crossed.
This is not a view of social organization born out of fixed binary divisions,
civilized versus tribal, white versus not, friend versus foe, universal versus
particular, normal versus abnormal, and so on. The word "normal" is used
by this self-searching U.S. Marine in a nonnormative sense. He refers here
to the connection between any number of different cultural options, all
equally marginal or central depending on the mission situation. Anthropological open-endedness is thus a founding premise of network-centric
COIN theory, and the devil dog's appreciation for "Bedouin women's poetry." On the order of what the corps calls "Semper Gumby" (i.e., being
always faithful to bending and adapting like the famous amorphous clay
animation character from the 1960s), and fully indicative of the paradox of
normative human relations in the nonnormative sense, cultural fluidity is
the corps' way of becoming Gumby. The Marine Corps' Office of Diversity
Management is careful to keep systematic records of the racial, ethnic, and
linguistic attributes of enlisted members. In 2020, more than 40 percent
of the force identified as a member of a minority group.[109] But in the diversity management parlance the corps prefers, "All Marines are green,"[110]
as was Gumby himself (or to go the full distance of human objectification,
itself). Given the selective nature of this elite military branch, the attraction
toward a potential recruit lies not so much in having come from a specific
or marginalized culture as in arriving already open to multiple forms of
human difference so the new recruit is able to go green. To be made a Marine, as the Parris Island recruit depot saying goes, is to be remade into
something other than you are.

Similarly, as the U.S. Army's HTS program came to an end, fieldwork-
ers on the battlefield wanted to retain the ability to "improve the military's
ability to understand the highly complex local socio-cultural environment
in areas where they are deployed." The conflict here was over how to cordon off military targeting from "saving lives."[111] As a Wikileaks release in
2010 reveals, spying did occur within the HTS program through acquiring
information from the civilian members of the anthropological teams.[112] As
one veteran of the quagmires leading to the failure of COIN doctrine in
Vietnam suggests, the hunger for cultural knowledge during the high point
of the HTS program was bound to be unsatisfactory for fighting the insur-

gents of Iraq and Afghanistan. As long as COIN operatives kept "anthropo-logical problems" too far away from "technology-based" solutions,[113] and as long as "fieldwork" was based on "non-material research",[114] it would be hard for war planners to know what to do with the knowledge of culture.[115] HTS's failure resulted as a problem of "information management."[116] Les-sons learned about the use of anthropology in battle emphasize, rightly, as Kevin Golinghorst puts it in *Mapping the Human Terrain in Afghanistan,* that "the nation's collective resources have yet to fully understand the di-versity of the people let alone assemble and display the many ethnographic layers of this nation" (1). Without a census conducted in Afghanistan since 1979, key words such as "big picture," "overview," "vastness," and "variety" suggest a failed account of Afghani culture "as elusive and as fragmented as the people themselves" (1). Moreover, four years into the HTS program, "the consensus is clear that a comprehensible understanding of culture and the human terrain is necessary but the means to get there seems to be still in dispute" (26). The word "means" in this sentence is telling. Even in the case where human and terrain are mutually embedded, what is still miss-ing from the HTS program is the technology capable of sorting through excessively large scales of anthropological difference. What is missing is neither the human nor the terrain part of the HTS program, but the other key word: system.

John Stanton has thus collected two hundred pages worth of journal-istic and primary sources tracing the course of the HTS program's prog-ress between 2008 and 2013, referring to it as "the program from hell."[117] His introduction to a volume on the subject marks a continued division between "social mapping and technology development," citing remarks about the program's founder, Montgomery McFate, as "a poisonous in-dividual," "the crazy aunt in the room," and a "hustler."[118] But in a profile of McFate in *Elle,* a glossy fashion magazine with an appropriately global audience, "the 42-year-old cultural anthropologist is a paragon of the 'American Spirit.' She is a liberal Democrat who initially opposed the war in Iraq but is now in the thick of it. . . . Oh, and she consults the *I Ching* before making career decisions."[119] In her response to the AAA's objection to using anthropology for human terrain purposes, and to other critics of the HTS program, McFate focuses on overcoming the "intellectual isola-tionism towards the military" within academia, if not also within society at large.[120] Perhaps more damaging, she suggests that the AAA's desire to protect anthropology from war amounted to "epistemological censorship," as well as "supreme naivety about the nature of knowledge production and distribution." McFate continues, "it is in the nature of knowledge to escape the bonds of its creator."[121] The debate between McFate and González, for and against the Human Terrain System program respectively, keeps the conversation squarely within the consideration of ethics and how best to either exploit human relations or presume to protect them. However, even in their antagonism, AAA dissenters and Human Terrain System advocates have agreed to open the black box of politics as embedded in the study of human relationships. Neither party argues that there is such a thing as the production of knowledge in a neutral way, just like early depictions of object relations serve as target alerts in the stochastic knowledge experiments within CE-COIN. The common way in which epistemic nonneutrality is affirmed creates here a kind of political impasse. McFate celebrates the discipline of anthropology as "escaping the bonds of its creator" (22). But the escapee comes fully back to HTS program territory when the war-resisting anthropologist decries scientific objectivity in the study of culture. What HTS program discussants have overlooked on all sides is a conceptual shift in military technology. This shift presents less a cultural problem of human intersubjectivity than a reinvention of the species along computational lines. The turn to computer intelligence, along with "linking human factors to the physical terrain . . . with ecological analysis," is how "the HTS program will likely continue to evolve and grow in the future."[122] Moving away from Mead's living techniques of the oral interview, and similarly away from the scholarly monograph, this evolution jettisons language in the traditional sense for "visualizing socio-cultural terrain."[123]

Though the Marine Corps did not itself have Human Terrain System teams, as did the U.S. Army, it did use academic anthropologists as "cultural advisors."[124] There, too, the problem was less having access to too little data than of having too large a quantity of cultural information to put in written terms: "losing [culture] in the translation" between numbers and language.[125] The biggest obstruction to the cultural turn in COIN theory has been that there appeared to be no "there" there. In practice, HTS practitioners tried with minimal success to objectify subjective factors within insurgent groups. This was because subjectivity as such did not seem to exist in reliably mappable ways. But this did not keep war anthropologists from trying. One description of a *Marine Corps Battle Staff Training* facility emphasizes a telltale shift in the technical means by which a host nation's population is divided: "enormous video screens, a colorful map showing mountains, rivers, roads, rail lines, and ports, a second map showing the same landscape but overlaid with strange symbols—colored rectangles with X's and circles inside, black lines connecting the rectangles, and big blue arrows emanating from boxes, computer printouts providing a dizzying array of information, anything from the locations of police stations, to relationships between known insurgent groups."[126] The move from exclusively cultural to geophysical mapping does not supersede subjective relationships so much as it seeks a way to enumerate them within an overlay of human and material vectors. Moreover, the amount of information—now numerical, and therefore, no longer separated by subjective versus objective qualities—may be "dizzying" to the human observer. But it is legible to the machine.

Anthropology's eventual withdrawal from the HTS program was both an ethical decision to resist the targeting of human bodies and the long-standing disciplinary desire to resist quantitative knowledge work. But computation in war puts into play new forms of complexity, revealing as much about the physical world as the term "culture" tends to hide. When McNamara lamented about not being able to find a calculus sufficiently capable of measuring psychological warfare, he was hoping to signal a turn from psychic affairs to more materially data-driven ones. In this sense, McNamara wanted, without quite knowing it, a way to move beyond a subjective way of measuring human relations as they were entangled with corporeal attributes, the human being displaced by the human thing, as in the term "human terrain."

Prospects on this order are precisely what are offered up as the official tactics of asymmetrical war doctrine in the most recent *Counterinsurgency Field Manual.*[127] Drawn from the lessons of Napoleonic Spain, French Algiers, and British Malaysia, the *Manual* is both a tactical bluebook against urban insurgency and a sustained effort in critical race studies as war by other means. With its high dose of up-to-the-minute social network theory, this is a *post*-post-colonial studies text: its purpose is to wage low-intensity conflict within what is loosely called the "global civil societies" of our "host countries," and make "culture" a decisive "area of war operation." COIN in this manner focuses on those "paramilitary, political, economic, psychological, and civic actions that are necessary to defeat insurgency" (xxiii), and to place cultural areas of war operation within the charge of U.S. war tactics. COIN operations are thus designed to manipulate "identity-focused insurgencies" (24) as softly reinstrumentalized mechanisms of population control without presuming the separation between human relationships inside and outside of war.

Such strategies therefore "include determining who lives in what areas and what they do. Establishing control," we see once again, depends upon "conducting a census and issuing identification cards" (24). COIN theorists, like Vietnam veteran David Galula, repeatedly emphasize the importance of using an occasional census in insurgent areas as a form of population control.[128] The difference between classic COIN on the order of Galula and the stochastic counting offered in the newer CE-COIN is in the latter instance census work occurs in real time, and at every second of every day. The focus of the Everywhere War, to borrow Derek Gregory's illuminating term, is in the engagement of nonstate groups.[129] Insurgent threats exist along "the widest possible spectrum," and are based on "profound cultural and demographic tensions that blur traditional categories of conflict."[130] As the practical extension of COIN doctrine dating back to Vietnam, we can also note that the HTS program seeks to "define the population as an operational environment . . . as defined by ethnographic data and non-geophysical information . . . that may be referenced geospatially, relationally, and temporally."[131] The overlay between geophysics and ethnicity maintains a certain degree of difference between human relationships and material ones, but the difference is not absolute. The word geospatial signifies a special kind of communal association in this version of war anthropology. "Population as an operational environment" targets immaterial forms of cultural belonging but makes culture visible. Its visualization exists in the same way war planners plot more traditional forms of geological terrain.

To cite the language of Overwatch, designers of the MAP-HT tool kit for Human Terrain program mapping: "Social network analysis and real-time collaboration may be used to collect, store, and process, analyze, visualize, and share green data through all phases of the civil information management process."[132] However, early evaluations of ethnographic mapping on the order of what HTS program parlance calls green data were less successful than its mappers would have wished. As HTS team members testified in 2009, "MAP-HT is still far from being fielded," and "MAP-HT has never worked."[133] This attempt to visualize culture for the purposes of war is indicative of military goals for bringing telepresence to more traditional forms of conceiving the battlefield. While the desire to geospatialize human relationships on the computer screen was never successfully fielded in actual battle, the quantitative applications of war have nevertheless supplanted the earlier qualitative ones. In the wake of HTS 1.0, the "majority of military strategists agree that COIN's conceptual underpinnings are weak."[134] But defenders against a telepresent enemy also remark, "COIN is dead—long live COIN."[135] The goal for a revised and technically super HTS 2.0 holds on to the premise of mapping human relations as terrain. As has been reported, the army misled Congress and taxpayers when it claimed to end the Human Terrain System program in 2014. HTS remained alive in experimental versions into 2016 and retained a budget of about $1.2 million per year (bringing the 2007–16 total to $725 million).[136] HTS program efforts are poised to be reinvented and expanded, with new mapping tools, such as the army's Global Cultural Network indicates. A U.S. Army War College document is revealing about what future counterinsurgency practices will look like when it says, paraphrasing Joseph Stalin, "quantity has a quality all its own."[137]

The problem of mapping culture in the first iterations of the Human Terrain System program presented issues with the integrity of the data. But these issues were problematic only because the thing being measured was still locked into a zone of inaccessibility, or of being *no*-thing at all. The enemy-in-everyone had not yet been televised. But the resistance of culturalists to turn the humanity being into data does not mean that the data are not there. The turn in war anthropology from a focus on culture to geophysical notions of community in a new space and time makes it possible for the advent of posthuman COIN. The HTS program was supposed to map self- and group-identification in an ever-expanding, open-source data cloud, available to battle commanders in real time. Its goal was to produce a never-ending census map, or a computational tool for measuring communal ties that were mutable and fluid. While the practical application of the program fell short, its theoretical implications for reviving COIN theory remain relevant and clear. The term human terrain ultimately seeks to displace culture as relevant to war and, through new forms of computation, embrace a concept of humanity determined by things. Already we have identified the U.S. Marine Corps' embrace of Semper Gumby, accepting as its unlikely mascot the rubbery, green, humanoid Deviled Egg Head—a person turned into an all-inclusive network of things. Similarly, in the war-planning room, Green Cells map sewer lines, rivers, and insurgents, as well as the constantly shifting identities and allegiances that rise and disappear between population groups. This technique both flattens the elements of the battlespace (they are all equally a matter of data) and expands them (they can be recombined and overlapped as systems). There is only one requirement: the cultural and physical landscape must be processed in computable form.

In his famous reinvention of COIN doctrine in 2006, General Petraeus insisted: "Cultural awareness is a force *multiplier*" (emphasis added).[138] This is ground we have already covered, and it runs a consistent line through the earliest applications of war anthropology in the removal of Native Americans from indigenous land all the way through McNamara's lament about the failure of the body count in Vietnam. But with HTS program's term "white Afghans" in mind, we may see something else in Petraeus's embrace of culture: the term "multiplier" here is as significant as the term "force." This is because multiplicity continues to define counterinsurgency both in the ways insurgents use local populations and in the way computational machines are used to track and destroy them. As one COIN document puts it in 2014, "'The Green Cell' is a commander's planning tool to help better understand the environment he is in with respect to the indigenous population."[139] But as the doctrine is careful to add, this "environment" represents a form of knowledge writ larger than merely human populations (139). Green Cells add massive and diverse sets of data points to the traditional battlefield, recombining human and nonhuman relations as telepresent with one another. The document continues, "'Green Cells' add another category of actors to the equation" (139). But in this expanded sense of environment, nonhuman agency plays a role coequal with the cultural kind: "We Marines think about everything in terms of terrain" (141). If everything can be turned into terrain, then it can also be quantified. From there it can be weaponized as part of the battlefield.

The Canadian military has also initiated a Human Terrain System approach to war, which puts a new twist on the digitization of "identity drivers."[140] In 2008 they deployed what was called "white situation awareness teams." In line with the HTS program's emphasis on visualization, the team's objective was to "map out the movers and shakers of Kandahar, and reveal how they relate to one another."[141] The term "whiteness" as an explicitly tactical and situational problem is revealing. In the Canadian HTS team's use of white identity, whiteness is both deracialized and rematerialized as a coding application for counterinsurgency goals. In military parlance, red means foe, while blue means friend. Whiteness along this spectrum exists in a kind of grey zone: white Afghan designates whatever ethnic identity happens in the moment to be armed. Along the color lines organizing Afghan coalition forces, the worse kind of military violence is Green on Blue, meaning attacks by the Afghan police on American and NATO troops. But an insurgent may slip in and out of whiteness from one moment to the next, depending on what the battlespace presents.

Pashtun tribal areas in East Afghanistan and Northwest Pakistan might put into whiteness (as with the *spin gund*) or "white faction" (against the *tor gund*) or "black faction," so as to trace ethno-political divisions and alliances.[142] But this association between whiteness and war does not make sense in the Western way of conceptualizing race. White Afghans in COIN doctrine's terms are subdivided into multiple forms of whiteness. There is the whiteness of not being armed, and then there is the form of whiteness divided from the black factions as the terms "black" and "white" are used among tribal Pashtuns. White relationships visualized through Green Cells reveal a temporary and conditional belonging determined by military goals. The white Afghan is brought into existence through computational processes designed to treat friend and foe alike, or at least, to see how potential targets may weave in and out of each of those categories. In that sense, whiteness is a product of information management. If you have enough green in the data, you can move people in and out of whiteness by adjusting your operating parameters. As the Afghan whitens—fails to become white, or travels, day-to-day, in and out of whiteness—such knowledge can be recorded, transmitted, stored, and manipulated immediately, or at least as quickly as the data may be processed.

The unmooring and reattachment of whiteness in HTS-style warfare is consistent with COIN theory's updated version of race and ethnicity as relational and ever-changing. Alongside its heavy use of the term "culture," the 2006 COIN *Manual* relies on the notion of racial heterogeneity. It is an utterly inclusive document on the question of human diversity, even while diversity becomes an instrument of war: "no society is homogenous," the *Manual* reads (85); and while race is a key theoretical factor for COIN operations, "there is no such thing as 'race' to account for human beings; race is a social category" (86). Commanders are instructed to presume a network-centric basis for compiling data for different groups of races, and to manipulate as many varieties of racial belonging as may exist or be encouraged to exist in any given area of operations. The production of the white Afghan is a case in point of such a practice.

Culturally embedded war is thus a keystone of "global civil society" where new ethnographic technologies can be applied to "create divisions between movement leaders and the mass base" (181). The *Manual* continues, Commanders should "seek . . . cleavages between groups . . . cross-cutting ties between them; reinforcing or widening their seams" (87). As HTS literature further points out, the "cultural analyst" should be able to "gather, store, manipulate, and provide cultural data from hundreds of categories in a way that reaches back to US academic sources."[143] More data beg for different categories, and the more in this case produces white Afghans. In this rendition of whiteness, we have moved from biological notions of racial essentialism toward a form of color coding more adept to a form of counterinsurgency that is inclusive of human and nonhuman terrain. HTS doctrine thus continues, "The dynamics of identity politics and group loyalties are fluid, opaque, and variable across localities that counterinsurgents cannot afford to neglect [as part of] their legitimacy-building tool kit."[144]

The *Manual* defines globalization as nothing less than a global census event turned paradoxically against whatever residual protection it might afford civilian populations. The process of decivilianization occurs by affirming racial fluidity, revealing a disturbingly innovative aspect of post-human COIN doctrine. This affirmation is consistent with the softening of other divisions, those between big armies, between friend and foe, and more broadly, the overlapping of human and nonhuman agency. On this point, contrast the 2006 *Manual* with a 1943 U.S. Army pamphlet, *Instructions for American Serviceman in Iraq during World War II*.[145] This pocket-sized document was written to inform U.S. troops how best to assist the British in Iraq against Nazi infiltration, a different moment in the history of whiteness, to be sure. In the section "Differences? Of Course!," the soldier reads: "Sure … there are differences galore. But what of it? You aren't going to Iraq to change the Iraqis. Just the opposite. We're fighting this war to live and let live.'"[146] Similarly, the U.S. Marine Corps's *Small Wars Manual* for 1940 emphasizes the "lack of exact information about foreign people," and that there is no point for "information" of this sort. "Foreign people to the average North American are practically beyond understanding."[147] The further development of war anthropology would find new interest in foreign people once the United States entered the war. But in 1942, American policy broke with formal neutrality, overturning the principles laid out in President Roosevelt's aptly named "Quarantine Speech." As that title suggests, demarcations between national identities were clear, and the United States presumed it could be reliably cordoned off from "the epidemic of world lawlessness."[148] Far from the live and let live notion such political isolation once implied, posthuman COIN technologies open up access to the human being per se across national front lines. This access portends a very close proximity between war and ordinary life.

HTS research represents only the latest development in a long-standing interest in what we might call the militarization of space-time, surmising new dimensions of the battlefield by mixing quantified and qualified elements of war. As the 2006 *Counterinsurgency Field Manual* suggests, the soldier-ethnographer's task is to redraw global systems of "racial belonging though they may not conform to historical facts, or may drastically simplify them" (93). The keywords here are "historical" and "facts." The word "facts" belies subjectivist notions of culture as the determinate force in the creation of human relationships, a position the HTS program has struggled to visualize as data, given culture's empirical elusiveness. Alongside this elusiveness, the word "history" intimates a fluid space-*time* relation where fluidity remains at the center of counterinsurgency. Time itself becomes fluid once the HTS operator admits racial belonging is mutable and can contradict group identification as it existed before. This register of time is consistent with the way identities are made mappable as shifting spatial networks, rather than as conscious or consistent modes of affiliation. Identities change over time, and like the white Afghan, those changes are more apparent to Green Cell data analysts than they are to the persons being analyzed. This is because the frame of reference within which friend–foe identities are made apparent exist in a scaled-up and virtual way. As informational rather than merely subjective entities, insurgents can be targeted even before they know they exist as a threat.

The Human Terrain System program and the 2006 *Counterinsurgency Manual* commonly highlight identity-focused strategies as crucial in contemporary war. But the failures noted with mapping ethnographic relationships in the first stages of HTS program deployment had to do with its hunger for data. Commanders complained that the work was often "very"—as in *too*—"time consuming" to be of tactical use.[149] As one of the program's veterans noted: "It is generic training. Everything is extremely rushed, in part because they are trying to ramp it up so fast."[150] The conflict between scale (too much data) and category (genre training) is intertwined with a frustration about time (we were too slow). The MAP-HT application as it existed in past versions of the operating system failed to capture events in real time. In the case of Project Camelot, the most ambitious social science research project of the Vietnam era, Special Operations Research Office documents emphasize counterinsurgency as a particularly time-sensitive practice: "We must develop a social systems model which would make it feasible to predict and influence politically important aspects of social change."[151]

More capable forms of computational skill are needed to match the spatial and temporal complexity that so-called culture brings to war. This is not to say that qualitative expression is insignificant in the context of HTS mapping. It is not a matter of choosing between numbers or narrative exclusively, just like it is not a matter of choosing matter over human beings. Rather, the more challenging point is to see how the first term in each of these pairings transforms the second. They do so, to cite the U.S. Army's *Human Dimension Concept* protocols, in computational ways—with "access to social media"—by controlling "the rising *velocity* of human interaction" (emphasis added).[152] Described by the U.S. deputy chief of staff as a "major revision" in extant "human dimension initiatives," culture processed in faster time emphasizes "medical and physical" activities, rather than only "cognitive and social" ones. Here, too, communal relations and terrain commingle as "integrated components" of war (5). While the importance of "diversity" is also touted in the *Human Dimension Concept* document (*HDC* 15), the effect of putting the mental and the material aspects of fighting insurgency together is another example of the human being reconceived as terrain.

Whatever is left of culturalism in posthuman war anthropology is expressed in a technical lexicon. As the *Counterinsurgency Field Manual* explains regarding storytelling, here in stark but accurate terms: "The central mechanism through which ideologies are expressed and absorbed is the narrative. A narrative is an organizational scheme expressed in story form. Narratives are central to the representation of identity, particularly the collective identity of groups" (76). As the word "mechanism" suggests, the success or failure with which identity and collectivity cohere is a matter of media efficacy. Here the media at work is as much the storyteller's voice as the listener's machine. As a mapper of organizational schemes, the counterinsurgent's task is to intervene in the relationship between identity and collectivity. Here narrative becomes data, meaning a story processed in a tactically useful way or the rewriting of a story as it may connect to a massive array of other stories, not only through words or individual consciousness, but also through numbers with machines. With the same goal of scaling up the story by turning to data as presented in the 2006 *Counterinsurgency Field Manual*, military research funded by the RAND Corporation seeks expertise in interpretation. The Global War on Terror must capitalize on preemptive counterinsurgency actions—call this being receptive to stories told at accelerated speeds. Security, the document remarks, must engage with "*inspirational* rather than *operational* violence" (emphasis in original). The new COIN doctrine therefore "demands intervention *before* an attack occurs."[153] This time signature of responding to a hostile action before it takes place is apposite to the notion of war on a planetary and virtual scale.

An exposé by the geographer Stephen Graham titled "US Military vs. Global Southern Cities" reveals page after page of Defense Department documents outlining the temporal adjustments inherent to the future of (or better, the future *as*) war.[154] Such conflict, Graham observes, will be borne out of the ways in which "global southern" cities are being targeted full-time by military satellites as locations of "urbanized insurgence."[155] Using unmanned aerial video surveillance either deeply embedded within city architecture or loitering in the atmosphere over potentially hostile areas, computer software profiles normal movement patterns in order to detect anomalies in the microgeographies of the planet's civilian populations. Drone theorist Grégoire Chamayou introduces the rich concept of copresence to describe this state of permanent and invisible war.[156] Copresence means being right here and right now, in one's own time and place, conscious of one's daily activities and ordinary routines, but at the same time existing within the data-intensive world of security surveillance permanently on the brink of military violence. Consistent with the human being becoming terrain, copresence on this order "does not presume being conscious of co-presence," since the target/citizen is occupying two places at once: an algorithmic and virtual place and, simultaneously, a geographical and physical one.[157] But both ways of being present are equally (and fatally) real. This is—paradoxically—the close-up *and* far-away nature of drone vision.

Copresence as identified by Chamayou depends on wide-scale pattern recognition capabilities or, as we have been using the terms, virtual ways of sussing out what is there. Human beings create traces of movement every day and at every moment. Such movements are invisible to our awareness, given not just our ocular but also our memory limitations. But even as virtual patterns, our movements are empirically there. They just need the help of better calculators than human senses can provide to trigger Graham's drones into dispersing their payloads. By way of copresence, killing from a distance paradoxically pulls war within the normal activities of civilian as well as noncivilian life. This is a spatial matter. Temporally, the ability to target an insurgent before the offending act occurs shows how speed moves military violence closer to humanity than before. These new capabilities "compress the kill chain" according to a "first look, first feed, first kill" operation, as the Raytheon Corporation puts it.[158]

Such video-based technology compresses the time it takes to find a target with the time it takes to kill it, moving vision itself as close as possible to violence: "Before you can drop your weapon and run," the arms-sales pitch reads, "you're probably already dead."[159] Preemptive war technology on this order effaces the enemy even before that enemy is designated as such. The target itself is not seen and then destroyed: seeing is its destruction. Real-time telecommunication in drone war offers no duration between violence and representation, no peaceable distinction between the time it takes to take a picture and the time it takes to strike what is in the camera's frame. The application of posthuman technologies required to render targets visible within a new register of space and time marks a move not so much away from the HTS program. Rather, drone vision exemplifies the territorialization of atmospheric space. But as we will see in the following chapter, the computerized battlefield is not yet fully expanded. Posthuman war reaches inward, as much as it has out and up. Thus in the final chapter of this book, we will explain how military violence invades the human brain.

3. War Neuroscience

We have based our authority to detain not on conclusory labels, like "enemy combatant," but on whether the factual record in the particular case meets the legal standard. This can be demonstrated by relevant evidence of formal or functional membership, which may include an oath of loyalty, training with al-Qaeda, or taking positions with enemy forces. Often these factors operate in combination.

—Harold Hongju Koh

THE FUNCTIONAL COMBATANT

What combination of "form" and function" are necessary to make military action conform to the law? Harold Hongju Koh, legal advisor to the U.S. Department of State, disagrees with the Red Cross's position on detaining suspected insurgents. He is explicitly rejecting "conclusory labels, like 'enemy combatant.'"[1] In Koh's preferred lexicon, such restrictive classifications are no longer needed because they impinge on the state's ability to locate the opponent in war. What matters in the crossover between civilian-suspect and enemy-combatant is a combination of factors where the friend–foe distinction is provisional, temporary, and mixed. Koh thus makes a key distinction, targeting those who are functionally—rather than only formally—engaged in war. Fact is determined situationally, according to one's position, through associations stretching the legal conclusions of the Red Cross. If you are positioned within a certain proximity to hostile agents (who are also on the move), and whether your formal intentions are explicit or not, you may be designated as part of the enemy forces. Position trumps form in the same way that demographic fluidity in the U.S. census, and the weaponization of culture in the Human Terrain System program, reject race and ethnic essentialism. Selfhood within the too-narrow confines of mere self-description is displaced once again by a notion of identification dispersed among the widest possible array of nonsubjective actions and effects. The oppositions between traditional categories—self and other, subject and object, peace and war, civilian and soldier—give way to a more expansive mode of war operations.

Stephen Graham refers to expansion as dependent upon surveillant simulation. Here "automated surveillance systems are fed directly into dynamic facsimiles of the time-space 'reality' of geographic territories."[2] The key words here are "automated," "dynamic," "facsimiles," and most important, "reality." Drawing attention to the functional effectiveness of targeting enemies rather than merely formal ones, Graham defines virtual geography as an artificial softening of the friend–foe distinction. But this distinction is also verifiably—as in legally—real. This precarious form of social existence rejects any number of stable conclusory labels, as is indicated by Koh. In the midst of the ever-expanding conditions of war, legal advisors to the U.S. Department of State (there were 175 lawyers in 24 offices when Koh dismissed conclusory labels) struggle to parse the disappearing distinction between soldier and civilian. At the same time, what we take to be reality in the superficial context gives over to a new condition of existence where representation and reality are no longer functionally distinguishable. The civilian becomes a functional combatant, though not a formal one, which begs a more complicated set of questions: How do new media systems reorient who is in the theater of war? How do new media systems move between the physical realities of the battlefield and its virtual spaces? At issue in the move from form to function is how war flows freely through the body politic. But the term functional combatant also alludes to military violence working within a corporeal geography. At the same time that virtual media produces new battlefields, the distinction between body and mind disappears.

Koh calls for a new era of "translation or analogizing principles for the laws of war and international conflict."[3] The word "analogizing" in this sentence is precisely the right one. In the more obvious sense, to construct identity according to "analogy" highlights the radically relational nature of the citizen-subject, whom the state must now presume is combat ready, whether the enemy is conscious of such readiness or not. Political violence lurks within the form of the identity the state ostensibly wishes to protect; or better, violence becomes part-and-parcel of what it means to be protected as a citizen. In protecting the citizen, the law is paradoxically reconceived to presume the functional equivalent of insurgency in all places at all times. Such a highly figurative—but also entirely real—oscillation between form and function is another example of the boundaries between friend and foe withering away. Here, as in HTS applications, we see a remediation of identity through the techniques of mapping—figuring, not formalizing—and how this process of remediation depends upon a more finely tuned theory of identity than anthropology has previously given us. Analogizing, in the U.S. State Department's sense, is central to the rewriting of both the laws of war as well as a set of related issues having to do with what counts as war matériel.

The image-of-us becomes the substance-of-us when war enters the virtual plane—military action without the formal awareness of the civilians functionally identified as foes. In this sense, analogies calculate and categorize what actually exists. But they also produce a unique kind of existence. Analogies arrange and classify, bringing reality to the surface of perception in a virtual way. They make meaning from an infinite array of empirically available relationships, which is not to say analogies are not empirical or the reality that they create is not real. What is really there can be sorted and re-sorted differently over time depending on the tools used to do the sorting. This introduces a theoretical hypothesis about the power of postformalist (i.e., functionalist) ways of conceiving identity and social relationships as positions in Koh's weaponized sense of this term. There is political significance in the link between the integration of differences among identities—which is truly massive—and the law. Under the conditions of global terrorism, analogical thinking enables a more malleable distinction between citizen and combatant in the manner in which Koh describes.

But there is more to the picture than just the friend–enemy entanglement. Analogizing as a war application also mixes the animate and inanimate, material and mind, the real and virtual, if not also the living and the dead. Analogy in these ways can be considered part of posthuman war epistemology, central to the reconstitution of the human being per se. Mixing and mapping are headings under which we can further examine the adjoining of our bodies and our minds to the arsenals of the Everywhere War.[4] How do war strategists render the multitude of unseen movements on the battlefield visible, and how does making them visible expand war fighting into the human domain? In his discussion of the Western Front in World War I, Derek Gregory uses the term "corpography" to identify how the scene of battle called trench warfare was envisioned through the body of the soldier.[5] Under the conditions of contemporary war, we can now begin to consider the reverse: How is the body of the soldier itself envisioned as war terrain? How do the body's own functions, internally and externally, its biological systems, and its cognitive processes, connect to posthuman war?

Barbara Stafford defines analogy as a "practice for weaving discordant particulars into partial concordance."[6] Here she steers us away from the Kantian problem of the subject/object dualism. Instead, Stafford's definition of analogical thinking reveals a postformal approach, recombining parts and wholes, giving us a language to understand Koh's dismissal of conclusory labels. She helps us begin to explain the way function overrules form in the U.S. legal standards for supporting the global war on terror. The scientific grounding of analogy in Stafford's sense is therefore not a merely constructionist-based approach to knowledge. Her antiessentialism is empirical without being naively empiricist. It also leads directly to advancements in understanding how knowledge itself is produced, and from there how human cognition becomes embedded in military violence.

The importance of analogy for cognitive science has been recognized since the early 2000s. As Keith J. Holyoak, Dedre Gentner, and Boicho N. Kokinov put it in their introduction to the watershed volume *The Analogical Mind,* analogy "involves explicit relational matching."[7] It denotes "the ability to think in relational patterns" and to relate discrete "systems of connected relations from within one domain to another" (2–3). While they say "analogy lies at the core of human cognition," at the same time the editors are wise to leave open the possibility that other "primate species" (2) might be capable of analogical cognition as well. There is a key connection both to how categories are made, and how they are unmade. Holyoak et al.

rightly place emphasis on the productive nature of the media by which the mapping of identities occurs. Their emphasis on the means of production for identifying, expanding, and redrawing categorical boundaries suggests theirs is a function over form kind of argument. "As Greek and Roman civilization gave birth to Western science," they write, "analogy was enlisted as a tool for advancing . . . new kinds of systemic and empirically verifiable analysis" (5). The very nature of scientific progress is rooted in the tools used for "the generation of new categories and schema" (5).

Toward this end, analogy mediates the disassembly and reassembly of reality. If we take the word tools in a serious way, this process of mediation is not mere fantasy or wishful thinking. Instead, the matter of analogy becomes precisely that—*matter*—in the sense of finding certain correspondences between the things doing the representing and the things being represented. Mapmaking defined as producing correspondences between media, thing, and media-as-thing, at various degrees of proximity and scale, is of central importance to the theory of analogical cognition: "Mapping allows analogical *inferences* to be made about the target analog, thus creating new knowledge to fill gaps in understanding" (9, emphasis in original). Maps in this sense are mechanisms for projecting a virtual presence. If they are accurate, maps correspond to actual physical formations appearing or disappearing depending on the scope of the map.[8] The irreducible variability of mapping, the way terrain changes depending on any number of perspectives and according to boundaries of scale, is not anathema to the progress of scientific knowledge. To the contrary: "Analogy is *every-thing*," as Douglas Hofstadter writes (499, emphasis in original). Hofstadter emphasizes both the "every" and the "thing" in that provocative declaration. At stake in the analogical universe is not only how science progresses but also how war expands at the same rate our experience of reality does.

Stephan Besser would relate Hofstadter's "every-thing" concept to what he calls the "isomorphic imagination."[9] Bresser writes in "How Patterns Meet," "Cross mappings between different domains . . . must first be created" (422). His interest in neuroculture aside for a moment, the word creativity must be given its analogical due. The creative *morphing* at work here does not make the imaginative part of Bresser's definition of thinking any less congruous with the equally evocative, stunningly complex—and above all anthropologically *amorphous*—dynamics of the real world. Like the physicist David Deutsch's definition of good art, science produces a new sense of universality by "cross-mapping" and by maximizing the courses of "information flow" (438). The scientific imagination in its isomorphic sense finds relationships between disparate phenomena at new scales (both temporal and spatial), taking matter apart, and putting it together in new ways. This same infinitude of worldly difference joins the maps, the makers, and the tools together in significant ways for understanding post-human war. This is particularly true for surmising the limitless nature of military violence, which challenges the category of the human being, as in the case of Human Terrain System, as well as in war neuroscience. In an age where the tools for making knowledge are changing faster than we can evaluate their effects, we need to confront the isomorphic imagination as a posthuman condition at its ambivalent core.

Analogical mapping in this sense, and in the legal and political way that Koh refers to it, is an image-based affair. This is why one of the state's primary security tools is the algorithmically enhanced archive of the internet and CCTV. Analogizing in its functional mode is about geometric reason—mapping, once again—and is therefore simultaneously image- and matter-dependent. As a theory of cognition, analogy also has to do with the difficulty of distinguishing the difference between its hard and soft applications: mind and matter, the event and our knowledge of the event, the actual and the phantasmal, the visible and the invisible, the virtual and the real. This is why the concept of analogy must be put into a technical register appropriate for understanding how human cognition itself can become an instrument of war. Once again from Stafford, analogy designates an "interface where the processes of sensing and acting meet up with the intricate formal structures of the visible world."[10] Human and machine; the subjective and the objective; category, scale, and time—all of the key concepts of analogy—are active in Stafford's emphasis on the sensation and the action of the image. Sensation, writ here apropos to Deleuze as the experience of moving pictures, is an enigmatic formulation. In Deleuze's sense of contrasting the eye of the camera with the "too-immobile human eye," Stafford's concept of sensation involves affective and corporeal processes, mental and machinic techniques, blended into one another.[11]

Decades after Deleuze, neuroscientists commonly refer to human cognition as a process of "neural . . . super-computing," and as a "visual system of computation."[12] "Perceiving the brain as an information system, classical computer science reads its [the brain's] input continuously," paralleling two ways of processing data: biological and artificial.[13] The term for joining these computational systems gives rise to hyper-computation. A neuron is slower by six orders of magnitude, and relies on chemical gates, as opposed to the on-off switches within nonbiotic computer transistors. But analogical physical values can be assigned to neuronal "wetware,"[14] rendering wetware useful for warfare of a brand-new kind. The turn to neuroscience in war moves beyond the so-called cultural turn associated with the Human Terrain System program. Here culture morphs into a full and final realization of subjectivity as terrain where the brain itself becomes a war instrument. Matter, mind, and media converge. According to one Board of Army Science and Technology document, "tissue is material." The "opportunities in biotechnology for future army applications" begin with the manipulation of "biological things."[15] In the practice of posthuman war neuroscience, cognition is one of these things. Besser comments insightfully on David Johnson Thorton's idea of the New Brain, along these lines. As a computational entity, the brain "presents itself as an assemblage of networks and circuits and wiring diagrams" and, citing Thornton, "'a field of combinatorial interaction, too dynamic and dispersed to be bound by relations of correspondence' between structure and function."[16] What makes the New Brain *new* is this moving of form together with "function," as in Koh's conception of virtual targeting, only now at the level of thought. But at all levels—the juridical and the epistemic, the cultural and the biological—objects are determined in an isomorphic way. New visual technologies based in algorithmic routines map new battlefields without conclusory labels.

Functional military violence, on this order, not only revolutionizes how targets are identified but also creates new kinds of spatial arrangements. These can be both very large-scale ones, as in drone war, and very small ones, as in the weaponization of synaptic nodes. Human geography works in the case of the New Brain the same way it does on the networkcentric battlefield. In both instances, images render material elements, like objects, bodies, and terrain, out of immaterial ones, like subjects, minds, and cultures. Thus, neuroscientists ask the question: "How do the

computational features of the human brain influence the 'realism' of a virtual environment?"[17] They do so in a way that redefines reality itself, a kind
of rematerialization of certain actions—like thought—no longer limited to
subjective experience. But this rematerialization is also a new way to think
about cognition as involving calculation, and calculation as fundamental
to war. The realism at work in the human brain "depends on the quality of
the simulation that is fed back into the nervous system."[18] In this sense, the
virtually real is "computationally defined as a 'realism' function."[19] This is
an important aspect of Koh's doctrine of the functional combatant. But the
further-reaching point about computational realism is the way technology
opens the human body to posthuman war, going inward toward the brain,
rather than outward toward the terrain-ing of mere cultural relations.

Thorton's term Brain Culture is thus significant for marking a move
from the human terrain of social relationships in the theater of war to the
materiality of thought. So-called culture in Thorton's use of the term is the
reverse extension of Human Terrain System logic. Thus, the ad hoc Committee on Military Intelligence Methodology for Emergent Neural Science
proposes a new research agenda: "neurotechnologies as weapons."[20] It depends upon "advanced socio-cultural neuroscientific models of individual
group dynamics based upon theories of complexity within BMI [Brain
Machine Interface] frameworks."[21] The committee may have granted the
Gates Doctrine's premise of seeing identities as cultural constructs, but its
reconstruction is regarded as a way to get closer to the "node-edge of cognitive and behavioral interactions . . . which might be viable to identify specific targets for manipulation."[22] DARPA's initial investments in the integration of soldiers and machines as computationally connected began in 1974
under the Close-Coupled Man/Machine Systems program, later renamed
Biocybernetics.[23] The latest updating of this technology further shows how
social and cultural activities are detached from their subjectivist moorings.
A different kind of construction—the kind called engineering—is at work
here. The U.S. Army's twenty-first-century biotechnicians are encouraged
to focus on machines not merely as related to human cognition but also as
a way to turn biophysical reality into weaponizable substance. "Memory,
self-replicating systems for wound healing, shock therapeutics," become
part of the posthuman war arsenal. "The determination of target threat
molecules; proteins for radiation resistant electronics; and other kinds

of bioinspired materials,"[24] expand what constitutes war matériel. As remarked by the *Journal of Special Operations Medicine*: "Now, discoveries made in the exploration of human health through biotechnological methods can clarify the law of life and the molecular level, which makes it possible to regulate and control the functions of the human body by adjusting its ultra-micro structures to gain powers of defense and attack."[25]

Within a bioinspired military model, human cognition is recalibrated along new lines of complexity and scale. Those lines extend so far as to make the notion of culture somewhat outdated. The end of culture in this sense gives way to an affirmation of the machine, which is given as much agency as the individual subject whose mind becomes machinic, given a newfound compatibility with computational media. This compatibility emphasizes the body over the mind, the realm of means over ends (recalling Kant), and technology over purely human experience. A new law of life places artifice over—combines it with—the physical relations we may still call the real. The virtual nature of reality is the definition of advanced ways of interfacing with machines. Thus "computational biology," referring to "the psychological and physiological processes underlying human information processing," becomes "the bottom line . . . for future war fighting applications."[26] This is not the same thing as the use of body markers in the form of identity cards, which the Biometrics Identity Management Agency (BIMA) used to keep track of the population, for example, in Afghanistan during the Human Terrain System program.[27] As a premier organization of the DOD, BIMA is in one sense on the cutting edge of military identity science. This is so because BIMA has the potential to store myriad forms of corporeal variation, going well beyond the more restrictive ones of race, ethnicity, or gender. The DOD's biometric database can theoretically grow as large as needed to accommodate the singular kinds of physical variations occurring within each human individual—for example, with fingerprints or iris scans. But computational biology as applied to the human brain goes one step further than "battlefield forensics"[28] of this kind. It makes military hardware out of the software of the human body, rather than staying on the surface of it. Truly *superior* "identity superiority"[29] invades the human being and recasts neurological processes as a database waiting to be mined.

Making "the law of life at the molecular level" a systematic part of war denotes a functional equivalence between "defense and attack" in both the military and the medical senses of the term "life." The convergence between the Forever War and increased political access to the substance of human existence is present in the concept of "biosafety."[30] If the state's prerogative is to decide who is a friend and who is an enemy in times of war—as Carl Schmitt famously theorized—its prerogative has been extended to a form of dominion internal to the human body. On the network-centric battlefield, corporeal existence is not like a zone of insurgent warfare. Rather, the human body overlaps the battlefield as one more computational grid. By way of transition to a fuller discussion of contemporary war neuroscience as the reverse extension of the Human Terrain System's terrain-ing of culture, there are further philosophical issues to retrieve. Here we can recall an important encounter between William James and Henri Bergson. Together they challenged what James called "the intellectualist method" where subjectivity rules over substance, which is precisely what contemporary war neuroscientists are leaving behind.[31]

Both James and Bergson called into question an idealist approach to the material world, and did so by giving primacy to matter over mind. This is the first way a Bergsonian theory of mind foreshadows war neuroscience. According to James, Bergson was right to reject the idea that thought alone is "an adequate measure of what can or cannot be."[32] Even in such a simple phrase, the payoff in what James proposed beyond intellectualism is clear: it is a philosophical error to presume concepts exist independent from the interaction between nonhuman agencies. Such an error posits a false division between what is supposed to be known and the being who—with the right tools—is producing the knowledge.

As James explains further in his affirmation of Bergson's work, this false division can be overcome if we allow for the computational logic of neurological processes. This is the second way Bergson anticipates war neuroscience: the means designating our practices of measuring reality (recalling Deutsch) are always also part of the reality being measured. James is not saying that the measurement of reality and the real itself are seamless extensions or are simply corresponding parts of one another. Such a position would amount to realism's own latent idealism, and would fall into the correlationalist fallacy Meillassoux attributes to Kant. It would also make Bergson's realism vulnerable to flattening out all variations of the real into one superficial and undifferentiated substance. This is something Manuel DeLanda circumvents by introducing a concept of matter as "material-energetic-informational" and subject to "emergent patterns of forms . . . that are irreducible to their emergent properties."[33] What James is establishing, before DeLanda, is a materialist basis for consciousness exceeding, and to some degree determining, what the mind thinks it knows. Bergson's critique of intellectualism is therefore "revolutionary," according to James.[34] This is because his alternate line of realist philosophy advances beyond subjectivity. The unconscious, human nature, communicative reason, and so on, are seen as ineffective humanist ideals. Bergson explains the connection between substance and thought by recombining them in quantitative terms.

This is where Bergson's earlier brain theory comes in, and how it illuminates a third aspect of posthuman war neuroscience. Following the emphasis on matter and the productive, as well as referential, capacity for virtual-reality making tools, Bergson alludes to the importance of computation. James writes accordingly, "Many of his [Bergson's] ideas baffle me entirely. . . . I doubt whether anyone understands him *all over,* so to speak."[35] Without putting too fine a point on James's prose, I emphasize "all over," as James seems to with the curious words "so to speak," simply to emphasize alongside his other admirers something at the heart of Bergson's realist revolution: the issue of plurality, scale, and numbers. Pluralism at the molecular level is the problem a war-based law of life sets out to solve. Bergson presents a most capacious scale of *all-ness* in the most radical and "empirical" sense. This leads to a neuroscientific theory of cognition consistent with the terrain-*ing* of the human being in counterinsurgency theory. But unlike its culturalist approach to human geography, the all-ness at work in Bergsonian realism emphasizes a missing link between sense and matter. This is the same link exploited by biotechnical applications of war.

LIVING MATTER

For Bergson, consistent with the idea of human mapping but with a chemical and molecular twist, the brain is not simply a transmitter or receiver of images. Nor is our reference to reality delimited by processes enabled by language. His is neither a semiotically nor an anthropologically based theory of knowledge. Rather, it is a corporeal and a technical one. "That matter should be perceived without the help of a nervous system and without organs of sense," he writes, "is not inconceivable; but it would suit a phantom, not a living, and therefore, not an acting being."[36] The affirmation of living matter is explicit here. But we do not have to remain as baffled by it as James. If James is right, we cannot interpret Bergson's emphasis on life or sensation as reducing reality to mere mental content or to conceptual abstraction within this-or-that set of transcendental ethical choices. Again, this would just be another version of Meillassoux's idealist Kant and would affirm Kant's false separation between the categorical imperative of humanity (the realm of ends) and the tools with which the species makes sense of the world (the realm of means). Bergson explicitly objects to the notion of a sense-matter correlation qua intersubjectivity, objecting to this as a false parallelism between the matter *of* the minds and the matter *about which* minds are thinking. This objection hinges on the more challenging hypothesis: minds think about matter while themselves being part of matter, and the tools needed to understand this process can process scale, category, and time in new ways, mixing human and nonhuman forms of intelligence. The scale of materially quantifiable neuroconnections exceeds linguistic ones. In Bergson's notion of "psychical" life—a complicated term informing Bergson's later theory of war—the living is composed of variable interlocking connections between both animate and inanimate stuff.[37] Bergsonian scale goes in the direction of the real-world-as-more-than-myself. This means that life operates in quantitative terms: not simply who am I, or who is not me; but once again, to what degree am I other, and in relation to how many? In this sense, category is numbers dependent. Echoing Cantor vis-à-vis Badiou, objects are internally divided, knowable, or invisible within the sets that make them so. This is the sense in which reality is computationally driven. So too, it turns out, is thought; and so is war.

The title of Bergson's 1904 lecture before the International Congress of Philosophy at Geneva in 1904, which had particular appeal for James, was "Brain and Thought: A Philosophical Illusion."[38] The "and" in this title calls out the false "parallelism" between subjectivity and objectivity underwriting idealist philosophy. Bergson's "radical" alternative, as James puts it, presents a more subtle hypothesis. For James, the philosophical illusion at issue in Bergson's brain lecture is staged as a specific example of a larger confrontation between idealism and realism precisely on this order. For

Bergson, the philosophy of the brain becomes a way to work through this confrontation as a problem of scale. His realist theory of all-ness requires him to explore the importance of neurophysiological movement as a fundamental part of thought, and not just as incidental to his general theory of knowledge. Bergson is interested in modifications within bodies exceeding conscious perception. Though these modifications may supersede our conscious ideals, they nonetheless reveal a crucial connection between our perception of things and things themselves. Bergson offers this enlarged philosophical picture by joining energy and matter, two things we share with—because we are part of—the physical world.

To use Bergsonian insights to explain contemporary war neuroscience, we must give full credence to the way James frames his own appreciation of Bergson: first, James says, he "came to philosophy through the gateway of mathematics"; second, James comments, "the old antinomies of the infinite provided the irritant that first awoke his faculties from their dogmatic slumber."[39] The kind of mathematics worth having for the philosophical realist is the kind that takes seriously the ontological implications of the infinite. James takes the issue of scale as a mathematical pathway, leading away from the kind of intellectualist reduction of infinity as explainable according to pregiven conceptual abstractions. Realism-as-pluralism on this order is not the same thing as explaining the infinite away as ungraspable (it is incrementally graspable) or claiming to have accounted for infinity once and for all through permanently fixed *generic* omnipotence. Categories are forced to change (like disciplines) because new numbers of objects eventually exceed them. James says Bergson is better than the intellectualists for responding proactively to the infinity irritant because his realism accommodates matter at maximal scale. In contrast, subjectivism depends on reality reduction.

This ought to be enough from James to realize the importance of the "quantitative discreteness" at the core of Bergson's philosophy. But we should not yet leave behind what James calls those "intermediary conductors." This is because war is making new use of the brain mapped as just this kind of entity: electronic media whose conduction is both mapped by computation and is itself computational. Eugene Thacker thus defines biomedia in helpful ways for surmising the conduct of war by other-than-usual means: "a process in which a functioning, biological materiality self-manifests, caught in the poles of immediacy and hypermediacy, the 'body itself' and the body enframed by sets of discourses."[40] Only we would have to press on the term "discourse" to keep it from presuming to transcend the materialist underpinnings of biomedia as a scientific concept. What James is intimating about Bergsonian realism not only brings numbers to bear on matter but also prefigures biomedia as a process of mediation technically compatible with the reality being mediated. Idealism ignores mediation or assumes that subjects and objects line up in purely subjectivist ways. Thus, James says, when intellectualist approaches to knowledge "arrest our concepts so that they may be made congruent with ourselves," they are lessening the force of reality, which is always predicated on the irritant of all-ness. "Abstract concepts," he continues, are simply "moments dipped out from the stream of time. . . . Real time plays no part in [this kind of] calculation."[41] The distinction between idealist time and real time depends upon more and less satisfactory forms of biomediated calculation. James suggests we consider time as a matter of "flux," where the *flux*-ing is not merely human dependent. "Reality is too concrete to be manageable," James continues, "because the intellect"—and here he refers to Kant directly—"is itself obliged to deny, and persist in denying, that activities have any intelligible existence."[42] For James, the most objectionable part of intellectualist epistemology is its denial of intelligence itself, where intelligence is writ as exceeding habitual forms of human consciousness. More significantly within the Kantian tradition, human consciousness is kept within his realm of ends, not means.

There remains potential for being baffled, like James says of Bergson, on the issue of means and ends. But we can become un-baffled by focusing a little more on Bergson's interest in the brain-as-image-maker. For him, cerebral processes and motor activities are inseparable. He further suggests that what brings them together is the image. In the same way as rendering the functional combatant visible within the virtual-reality-rendering practices of war, the brain both maps reality, and is itself mappable, in revealing ways. For Bergson, nerve centers lend themselves to cartography, or more specifically, onto-cartography. This is evidence of a theory of cognition where thought exists with a determined relation to space: thought can be geometrically drawn. But again, like the functional combatant introduced by Koh, thought is mapped as a moving set of relations, not fixed ones. In this sense, the mind "takes snapshots," Bergson says; and we might add, these snapshots are shot again in the form of the neuroscientific brain image. The movement later, from snapshots to shooting in war, is only a few steps away. The duration between thinking about a target and firing a weapon speeds up exponentially because the two kinds of firing systems (the brain's and the bullet's) are brought to together in a compatible way. To the degree neurological images are fluid and move fast—they change with memory, and at the speed of electrical current—"the mechanism of our ordinary understanding is of the cinematographic kind."[43] This gets us very close to Deleuze's interest in what he calls fact-images.[44] Bergson's sense-cartographies depict arrangements of matter interfacing with, and being altered in relation to, real-world spatial arrangements at an infinitely large scale.

In the same way, the scale of the human connectome—defined as the totality of connections between the neurons in the nervous system—is both singular and exceedingly large. There are infinite combinations through which cells in the brain could potentially connect with each other. Cognition is therefore *mass* dependent in the sense of being both material and involving large-scale movements. So, too, is our ability to model synaptic connections. When a neural network is taught, it changes physical structure, or its state space: "Training a network ultimately consists of partitioning its state space into an appropriately configured volume, with the correct sub-volumes and divisions, such that the network can embody the desired cognitive function."[45] The training offered here for human wetware is the same as computational hardware: minds and machines share the same image-based way of finding "hyper-dimensional activation patterns."[46] Accordingly, the whole of this brain–image–matter arrangement contains movable parts that take in "certain movements from outside and turn them into internal movements of reactions inside."[47] The inside/outside dyad is a little misguided, however, because thought is already enmeshed within an ensemble of materially oriented space-time arrangements. This ensemble includes both the content that cognition processes and the computational means by which the brain turns the infinitude of physical reality into things that can be known.

The essential relationship between human cognition and the flow of data marks an important breakthrough in war neuroscience. The DOD-funded research agency DARPA has developed a five-year, $80 million research program aptly called Electrical Prescriptions, designed to regulate the peripheral nervous system artificially and without interruption.[48] Ultra-miniaturized wireless technology will be used to monitor the soldier's internal organs and regulate how they respond to injury or combat stress. The compatibility between artificially intelligent nanobots and the human neural network creates what researchers call a "closed loop system."[49] But the closed system creates an expanded entity comprising two systems: one human, the other machinic. The compatibility of the two systems is evident in the promise of a "minimally invasive insertion."[50] DARPA's microcomputers are injected through a needle as intelligent pacemakers for the purposes of controlling the soldier's organs. Like the brain, the soldier's organs function by way of signals moving back and forth between the human body and machines.

One DARPA-funded neurosurgeon, who is also a computer and electrical engineer, thus describes the brain as "the internet of the body," and as "a central processing unit."[51] "If only we could tap into it," he remarks. The reference to tapping conjures the language of close-quarter combat known as the double tap. The double tap refers to a shooting technique where two shots are fired in rapid succession at the same target (for training purposes, under 1.7 seconds), and within the same sight-picture. But the novel class of tappers used for tapping the human brain biomechanically is much more efficient. A new class of ultrafine units, two thousand times thinner than a human hair, called magnetoelectric nanoparticles (MENPs), can move via the bloodstream freely across the blood–brain barrier, permeating the brain's neural circuitry by the millions. Once they are positioned in the correct proximity to the human neuron, MENPs can send and receive information between human beings and computer monitors in the time it takes to have a thought. "Our brains are pretty much electrical engines," the DARPA researcher continues, "and what's so remarkable about MENPs is that they understand not only the language of electric fields but also that of magnetic fields."[52] A special helmet with magnetic transducers will be used to relay signals generated by these electromagnetic fields as battlefields. In the overlapping of these fields, the specialness of the MENP helmet reveals a more complex situation of a helmet within a helmet: one data processor called the human brain, helmeted by the skull, beneath the DARPA-issued shell that is necessary to facilitate cross-function with the MENPs. The overlay between human and artificial forms of data processing thus leads to "the intelligent swarm": controlling hundreds of fighting robots on land, sea, and air, simply by activating the soldier's thoughts as a parallel swarming event.[53]

It is perhaps a little provocative to quote another revolutionary thinker on the brain, who also favored realism over idealism, and was a near contemporary of Bergson and James: V. I. Lenin. But Lenin helps us move the discussion of the brain–matter–media connection toward further examples of military neuroscience. Above we saw how James connected otherness, actuality, and the many, linking Bergson's theory of the brain-as-image-processor to the quantitative measures of cognition now capitalized on by DARPA's thought-engineers. In writing against what he called empirio-criticism (another correlationist ruse), Lenin insists, one "errs in believing that the brain secretes thoughts in the same way as the liver secretes bile."[54] Here, like Bergson, Lenin cautions against an idealist collapse between sensory versus material data; and like James, he insists that real relations are not reducible to subjectivity in isolation from them. As he says in *Materialism and Empirio-Criticism,* Lenin is against a "'realism' bedraggled by positivism, and other muddleheads who oscillate between materialism and idealism" (54). So let us underscore Lenin's commitment to "things-in-themselves," a loaded phrase of course, which he uses, after Plekhanov (and versus Kant), to set up an empirical theory of realism without being reductively empiricist (versus Hume). Lenin's connection to Bergson, and the way their work provides a conceptual bridge to war neuroscience, lies in the hypothesis: "Sensation cannot exist without 'substance'" (14, 34).

But more than this, an exchange between the two philosophers reveals not only a shared emphasis on the primacy of matter for human cognition but also a common interest in virtual reality as related to a law of life based in war. Where for Lenin concepts are subordinate to labor, for Bergson the cinematographic mind also works to put things together. Thought, he says, is "the transformation of the energy of external excitation," where the "image of the external phenomenon" is neither a "fence, nor a wall, separating consciousness from the external world," as "professorial philosophy" would have it. "Image" is therefore "directly connected to the external world" (44). For "energy" we may read: the brain's electrochemical networks as network-centric battlefields; for "image," the mapped brain and its mapping as translated via the helmet of DARPA's swarming MNEPs. In the same way that there is no inside–outside pairing to designate mind from matter, the image does not denote a wall. By extension, the closed system referred to by DARPA's brain technicians is not a closure but a form of biomechanical expansion, a set of walls within walls comprising a stacked set of skull/helmet enclosures. Using Lenin's term, the "word" wall denotes a new protocol—an enabling limit—allowing cognition-as-data to be shared bioelectronically between human and nonhuman entities.[55] In this sense, the skull-as-helmet concept can be thought of geometrically as a small circle within a slightly larger enclosure. Each enclosure contains its own circuitries and operates according to discrete movements of electricity and matter. But the skull-as-helmet concept reveals connections between their interfacing common nodes, as the concentric circles move outward and in common toward the larger circle of the globe at war.

The Bergson–Lenin connection establishes a theory of cognition not just as media dependent—running counter to Kant's realm of ends—but also as a way of processing thought itself as media phenomena. Indeed, "mediation" is a leading term in contemporary neuroscience, as in the remark "the principal task of the brain is to mediate . . . the divide between the vital requirements of the body . . . and the ever-changing word around us."[56] But the point to remember is that the mediation and the matter being mediated are organized in compatible ways. The image theorized in cognitive science denotes a sensation interchangeably human, material, and computational. It is, as Bergson says, a mind–matter exchange existing in the common form of "motor articulations." This is one way to explain how a soldier's thoughts alone (although they are never alone) can be relayed into the kinetic movement of weapons on the battlefield. "Thought and consciousness," Lenin says (quoting Engels in *Anti-Düring*), "are products of the human brain"; and further, "the material, sensuously perceptible world to which we ourselves belong is the only reality" (82). "Our consciousness and thinking," he continues at his Bergsonian best, "however supra-sensuous they may seem, are the product of a material, bodily organ, the brain. . . . Mind itself is merely the highest product of matter" (82). Neither Lenin nor Engels shied away from joining the terms sense and material, or mind and matter. Moreover, their revolutionary epistemology expands the concept of production to reveal the essential relation between thought and work. In this way, Lenin gives back to the brain its labor power and offers philosophy both its technical and its corporeal due. Bergson, too, refers time and again to the physical work" it takes to make knowledge. They could not know in advance how a materialist theory of human cognition would be turned toward the purposes of posthuman war.

CARTOGRAPHY AND VIRTUAL REALITY

Lenin and Bergson both subscribe to a theory of knowledge where, as Lenin says, "sensation, thought, and consciousness are the supreme product of matter organized in a particular way" (48).[57] Indeed, in *Mind and Memory* Bergson states that matter is "an aggregate of images," which exist "halfway between the 'thing' and 'representation'" (9). The image is simultaneously corporeal, to the extent that it modifies the body through "the afferent nerves of the cerebro-spinal system" (21). This schematic treats the brain and the objects affecting it as overlapping operating systems. "That image which I call my body" (21), Bergson writes, is both in and of the physical world. Corporeal existence is part and parcel of what he calls a "system of images" (9). James's interest in Bergson-the-would-be-mathematician does not easily spring to mind when we invoke the name of Lenin. But James's treatment of infinity in Bergson turns into Lenin's masses as theorized by Marx. Insofar as the two multitudes share a kinship, one epistemic, one based on labor power, science does its best to help us know more and act better:[58] mind is matter dependent. But a third "m"-word is crucial: the emphasis on image-based cognitive processes evokes a mind–matter–*media* scene. Media can be thought of here, on the order of cybernetics, in the way Wiener defines the gene as "a suitable nutritive medium of nucleic acids and amino acids" and as a "medium that lays itself down into other molecules . . . in an act of molecular multiplication."[59] For him, the body is "a system," or "a machine of a general sort."[60] Thus, too, in *The Fabric of Reality,* Deutsch defines biological life as a computational dynamic obeying "the same laws of physics as non-living molecules."[61] Genes, he says, "are computer programs . . . with complex control and feedback" mechanisms built in (171). "Life is really about knowledge as a physical quantity" (179). In an unintended blend of Bergson on the image and Lenin on labor, Deutsch continues: "Virtual reality is the physical manufacture of the rendered environment," a relation of production beginning and ending with what is "precisely real" (179).

For Bergson, too, "our actual existence . . . duplicates itself along with a virtual existence . . . as perception and memory."[62] Edward Casey is therefore right to remind us how Bergson's theory of memory is anti-Cartesian, and vastly different from Freud's limits of memory as the subjective domain of the psyche. (Notably, Bergson's *Mind and Matter* was published in 1896, the same year Freud coined the term "psychoanalysis.") Casey writes, "Bergson is convinced that memory images require reflection on the nature of matter."[63] As Bergson insists, the virtual nature of the image does not mean it is distinct from actual existence. Nor does virtuality mean a split between referent and representation. As Lenin describes in *Materialism and Empirio-Criticism,* media and mind both involve what he calls, with a distinctly Bergsonian inflection, "matter in motion" (166, 170, 176, 177).

For Lenin, image patterns are connected to a mode of production in the form of what he called practice. Summing up his theses at large, Lenin writes, "outside us, and independently of us, there exists objects, things, bodies and . . . our perceptions are images of the external world" (99). But insofar as "real being lies beyond . . . impressions and the ideas of man," it does not merely conform to "idealist 'coordination'" (108, 110); nor do "correct images" (109) simply "correspond" to (110) or "become identical" with the "subjective senses" (109, 111). Lenin goes on to great length in quoting Engels in the original German to oppose such "torturous theories of 'coordination,' and 'introjection,'" tagging the Kantian line of subject/object "correlation" with "idealist absurdity" (111). He thus draws attention to a key term from Engels: *stimmen mit,* translated as "to coincide with" (110). The contrast between Lenin's "coincidence" and Kant's "correlation" gives us a philosophical direction for understanding the skull-as-helmet concept, so important in the ability of war neuroscience for tapping into the kinetic nature of cognition: different information systems overlap in nonidentical, but compatible, ways. As Lenin insists, "Differentiation between true and false images is given by practice" (106). Here practice is defined both as a production and an act of reference capable of error correction. It is "the result of an action that proves the conformity *[Uebereinstimmung]* of our perceptions with the objective *[gegenständlich]* nature of things perceived" (106). Lenin is careful to spare us from taking the world as "'immediately given'" (107), while giving us both better (because more capaciously real) images, and better (because more truly equitable) worlds.

Given Lenin's term "immediacy" and his preference for mediation defined as productive work, we must add a point about temporality. This point runs along the same lines as Engels's refutation of the theory of *stimmen mit* and reveals the necessity of accounting for both time and space in Lenin's matter-as-movement-based theory of knowledge. In the place of "correspondence," Lenin prefers the word "coincidence," or even "concurrence" (111), as mentioned in reference to Kant. He purposefully offers a time-inflected word alongside a spatially oriented one for two reasons: one is to differentiate between objective realities and subjective idealism, the other is because he is interested in continued scientific discovery, which he links to the advancement of the natural sciences. Thus, for Lenin, "particulars [of matter] merge and dissolve in the conception of universal action and interaction" (156); and further, "real time is not the idea of time" (177). Given the massive increase in scale computation allows, we can see how "causes and effects are constantly changing places, and what is now or here an effect becomes there or then a cause and vice versa" (156). Here Lenin intertwines time and space by pairing the terms now/here and there/then. "For the materialist," he promises, "the world is richer, livelier, and more varied than it actually seems, for with each step in the development of science new aspects of science are discovered" (128).

Lenin's next sentence explains how the image can work with variation on this scientific order, not as a retreat from, but an advancement on, the real: "For the materialist, sensations are the images of the sole and ultimate objective reality." But here is the key point joining the "new" and the "more": Lenin means "ultimate objective reality . . . not in the sense that it has already been explored to the end, but in the sense that there is not and cannot be any other" (127). "Not any other" means not any other "objective reality." But it also means "not any other" kind of knowledge than the scientific kind. Thus, "real unity" is counter to the "pure Durhingian nonsense about . . . the fundamental homogeneity and connection of being" (175). Against what Engels calls "a few juggling phrases . . . [,] only a *long and protracted* development of philosophy and natural science" can detect the "idealist character" of the unity supplied by subjectivity alone (175, emphasis added). Image work disrupts the immediate connection between subject and object. It breaks habitual time, and forges new, more accurate, object-driven interconnections between reality and thought.

But time and again Lenin shows consistency with Bergson on the virtual nature of the real. He therefore prefers the term "image" over "symbol," or what he denounces as the "purely verbal." Like Bergson, he introduces a term more directly associated with the relationship between entities insofar as they change over space and time: a mapping word, one that introduces, if only by implication, not only "moving matter," but also the work of moving pictures as embedded in the mind–matter–media relation. "Matter," Lenin affirms, "is copied [in cognition], photographed and reflected by our sensations, while existing independently of them" (127). "To regard our sensations as images of the external word," he continues, "is to recognize objective truth, to hold the materialist theory of knowledge—these are all one and the same thing" (128). Lenin and Bergson break from idealism and affirm realism—by name—with a common set of theoretical questions. These questions get an empirical response only vis-à-vis the later use of still more imaging machines. I say "still more imaging machines" as if the mind–matter–media relationship were already resolved. But at least now we can ask the right questions about war neuroscience: To what degree can "sensation" be measured in terms of the "image"? How is mind–media–matter relationship mobilized in posthuman war? What kind of reality is the virtually real, and how does it link reality to military violence?

Deleuze more famously—and more directly than Lenin and Bergson—initiates these questions, when he states: we must designate "the cerebral process as object and motor of cinema . . . [,] something to do with scientific knowledge of the brain."[64] Here Deleuze introduces "the motor of cinema" as a mechanism of cognition and "the discovery of the topological cerebral space."[65] This is the sense also in which Levi R. Bryant uses the term "onto-cartography" as a form of mapping common to "the ontology of machines and media."[66] Keeping with a theory of knowledge as both movement of matter and the transmission of data between human beings and machines, the neuroengineer Tim Hanson calls the brain "three pounds of the most information-dense, structured, and self-structuring matter yet known."[67] He notes that the human brain runs on only 20 watts of power, whereas an equivalently powerful computer takes 24,000,000 watts. Similarly, in the Deleuzian sense, traceable back to Lenin and Bergson, and pointing forward to war neuroscience, thought and matter are commonly described as "signaletic material."[68]

THE HUMAN BRAIN AS IMAGE GENERATOR

The neuroscientist Antonio R. Damasio similarly offers a movie-in-the-brain theory of human cognition, where the brain is described as both an image-making and matter-arranging machine.[69] The brain, for Damasio, in *Descartes' Error,* is a "multi-media show" (115) he likens to "a large Cinema Scope screen."[70] Only, in correcting the metaphor of the brain as a "Cartesian theater" (94), Damasio theorizes brain activity from a distinctly Spinozist angle, adding both the operation of multiple cameras and the notion of thought as having its basis in matter. The mind is, apropos to Deleuze à la Bergson, both an image-based information processor and a mover of matter within specific instances of space and time. But the brain is not merely "a file of Polaroid pictures" (100). It does not present "rigid, facsimile representation" (102). To argue for pictures as such would be to offer what Deutsch, contrasting the mere snapshot with virtual reality, would call "image presentation."[71] And after Lenin, we must say again, realist explanations are both scientific and artificial.

Damasio thus clarifies his interest in the mind–media–matter relation: thought in his cinemascopic sense is transient. It consists of "topographically organized presentations" physically existing in what he calls "dispositional convergence zones."[72] Damasio's "topography" of mind is worth emphasizing for the way it reveals the interaction between the brain and other material surfaces, especially those relevant to applications of war. Other brain researchers also seek to quantify touch-feel perception. Here, touching an unusual surface is described as the interaction between multiple topographies. These are composed of the thing touched, the toucher, and the brain. Contrasting somewhat with the Human Terrain System program's weaponization of culture, a more literal weaponization of humanity is at work here.[73] In quantifying physical feeling, "the friction generated when a fingertip is stroked on a test specimen" shows a change in "tribological properties." These properties can be delineated in the form of physical changes in the brain's "synaptic grid."[74] The grid itself becomes visible vis-à-vis the Function Magnetic Resonance Image (fMRI) machine, in effect, producing an image of the ultimate image-maker, called the human brain. Functional changes in the brain's tissue at the level of microscopical substrates in one brain, and sometimes in more than one brain, produce the data necessary for the fMRI machine's own image productions.[75] In this sense, Damasio's term "dispositional convergence zones" repeats DARPA's skull-as-helmet proposition of overlapping, human and nonhuman, intelligence systems. As in the tempo of the battlefield, cognition has a crucial relationship with time. As Damasio asks: *When* are the synaptic cameras rolling, at what speed, and how do different brain regions fashion present and past in concert? This alludes to questions of scale magnitudes of order larger than human beings can understand without artificial means. The "trick of timing," Damasio says, is to get a "large scale system of neural activity . . . anatomically coordinated" so thought or memory can coherently occur.[76]

This is a very big job. If researchers at the Connectome Project are right, to trace the connection between every one of the approximately eighty-seven billion neurons in one human brain and represent their connection on a three-dimensional display, you would need about one trillion lines to successfully play connect-the-dots.[77] But the "autostereoscopic 3-D display technology" of your game board would still not account for brain plasticity (the way the brain's network can structurally change).[78] James was right to emphasize Bergson's interest in mathematics, since the brain-imaging technology behind war neuroscience depends on high-volume algorithmic computations. For Damasio, in *Descartes' Error,* cognition is a "physical and mathematical problem" (107). But in the context of one trillion-plus lines, it is also worth emphasizing Damasio's point about how "time does the binding" within the brain's "dispositional convergence zones" (105). Time "binds" the brain's spatial organization. Cognition can be mapped as a form of "topography in the visual cortices": when I see or remember "Aunt Maggie, I produce a discrete geography of her face" (105). But to produce Aunt Maggie's image in my mind, different synaptic zones will fire independently, sequentially, and/or synchronically, depending on the complexity of the role I give her in the organization of thought as a material and spatial arrangement. This is why Damasio describes the cinematic brain not merely as a picture presentation, but as a "visual and muscular" organ, where space and time are intertwined.

Damasio's version of the mind–media–matter relation brings us back to scaling up different systems and producing new and larger ones by achieving technical compatibility. For him, the brain itself is a "large scale system made up of multiple . . . noncontiguous . . . brain regions," which are in turn, "vastly more than the sum of their parts."[79] The "more" here not only designates the brain as a large-scale system but also portends openness to the transmission of data between itself and other systems. "Many images," Damasio continues, "might require explanation at the quantum level." Citing the mathematical physicist Roger Penrose, Deutsch alludes to "a possible role for quantum physics in the elucidation of the mind." This would require a theory of "consciousness . . . based on quantum phenomena occurring in microtubules—constituents of neurons and other cells."[80] Damasio's adjoining of cognition to quantum computing foreshadows the topic of a later scientific paper, titled the "Discovery of Quantum Vibrations in 'Microtubles' inside Brain Neurons."[81] Referring to the "'proto-conscious' quantum structure of reality," this research indicates how cognition is a physics problem, and one fundamentally embedded in mathematics. The term "spatial summation" referenced here is based on calculable frequencies, mapping the physical locations of electro-chemical signals as they happen, and calculating the various speeds at which the brain's substances move in relation to each other and to other moveable matter.

As this hypothesis suggests, the brain works in the form of moving "quantum bits, or qubits." In keeping with the idea of the brain as a kind of cinematic computer, visual awareness is cited as a particularly "favorable form of consciousness to study neurobiology."[82] Here the brain is characteristically defined as an "attentional mechanism, transiently binding together all those neurons whose activity is related to the relevant features of the single visual object." But note that vision is itself achieved in a material way, "by generating coherent semi-synchronous oscillations, probably in the 40–70 Hz range."[83] Vision becomes substance in this more literal than philosophical case. Images are rendered physically transmissible by transcranial magnetic stimulation (TMS) as electrical echoes recorded by the squiggly lines on an electroencephalogram (EEG). Certain algorithms are then used to calculate which echo-images are compressed and integrated in other parts of the brain, similar to the compression of digital photos into JPEG files. Similarly, as stated by Hameroff and Penrose in "Consciousness in the Universe," the qubits "orchestrated by Orch activity resulting in moments of conscious awareness or choice are related to the fundamentals of quantum mechanics in space-time geometry" (39). "The term quantum," allowing cognition to be mapped "geometrically," "refers to a discrete element of energy in a system where this energy is related to a fundamental frequency of its organization" (49).

But the equation of physical reality with consciousness apropos the quantum theory of cognition puts too much emphasis on the visual. Neuroscientific research also shows that there are dedicated receptors in the brain capable of accommodating thirty-three different bodily senses, rather than just the usual five of seeing, touching, smelling, tasting, and hearing.[84] The expansion of senses as such is only one of many ways to illuminate how the relations between one part of the physical universe and another are entangled: "Consciousness is an essential ingredient of physical laws" (40). Accordingly, the mind itself is shown to function in a network-centric way. The "perceptual moment" is described as "a series of discrete events, like sequential frames of a movie (modern film and video present 24 to 72 frames per second, 24 to 72 Hz)" (41). Lending further credence to the mind–media–matter hypothesis, "discrete voltage levels represent information units (e.g., bits) in silicon logic gates" (41). There are of course many uses to which this overlap between human and machine intelligence can be put. Nowhere is it being used more readily than in the military initiative linking war-fighting technology to the human brain. The significant aspect of this linkage for understanding posthuman war is that the terms "brain" and "technology" as I have just used them are inadequate to the stakes of what is being discovered here. The brain does not simply link to technology. It is technology, and it may now be extended by computational means into other technologies, linked almost seamlessly to war.

The U.S. Army Training and Doctrine Command's "Mad Scientist" initiative "explores the future operational environment and its military implications" along these lines, intermixing matter, media, and mind into the new "multi-domain battlefield."[85] In addition to developments in robotics, this research is clear about the all-encompassing reach that the army's "mad scientists" are looking to have: "All things in the future operational environment will be smart, connected and self-organizing; artificial intelligence (AI) and humans must coevolve, and human teaming with AI enablers will be the best instantiation of general intelligence supporting commanders in the battlefield."[86] The keywords here are "all things," "general intelligence," and "coevolve." "All" includes human cognition as well as the machinic kind. More revealing still is the symbiotic relation blending the two kinds of intelligence, allowing each other to expand beyond their former categorical boundaries. "Coevolution" may be read as achieving a higher form of military knowledge than either human or machine could achieve without tapping into each other's computational infrastructures. Within the kind of operations possible once the divide between cognition and artifice is removed, the "next fight will be characterized by electrons versus electrons."[87] They will flow without obstruction between a single expanded network of biotic and nonbiotic infohighways, pausing only against other electro-informational entities lurking somewhere in a battlefield encompassing "all."

Overlapping neuroscientific language with military vocabulary, the brain's "integrate-and-fire" mechanism works with computers, human brains, and more traditionally conceived forms of firing mechanisms. This commonality—our shared "perceptrons"—may refer to clusters of biological neurons and artificial neural networks alike. As an application of peace, Hameroff and Penrose connect their discovery of quantum Orch activity in the human brain to Eastern philosophy. "Buddhist writings," they point out, "quantify the frequency of conscious moments." The Hindu Sarvastivadins, for example, describe 6,480,000 moments of conscious thought occurring in the span of twenty-four hours (an average of one "moment" per 13.3 ms, 75 Hz). In similarly quantifiable and physical terms, some Chinese Buddhists suggest that a single thought takes place with an electronic value of one per 20 ms (50 Hz).[88] Hameroff and Penrose raise the prospect—from within a different tradition, and avoiding the overtly political interests of Kant—of a deeply personal kind of "perpetual peace," turning brain science toward an Eastern version of Enlightenment.

But Enlightened neuroscience easily switches over to the military kind. Jane Bennett's influential book *Vibrant Matter* is subtitled *A Political Ecology of Things*.[89] For good reason, she did not title her book *Violent Matter*. Yet military violence becomes part and parcel of what looks peaceful. Peace, or what military experts call "peace operations," is now a curious aspect of war. In Bennett's account, objects are most correctly viewed as *doing*—rather than simply *being*—things. In the same vein, Alfred North Whitehead, whose philosophy of science has influenced both Damasio and the Orch OR theorists, remarks on the "millions upon millions of centers of life in each animal body." Tellingly, he posits the scale of life on this expanded order in opposition to "Uncle Sam, who rules over and above all the U.S. citizens."[90] But Uncle Sam takes a great deal of interest in mobilizing the mind–media–matter relation as part of the network-centric battlefield. At the furthest reaches of the posthuman war arsenal, Whitehead's violent Uncle Sam and Damasio's peaceful Aunt Maggie are about to join hands in an unholy alliance. Further from the U.S. Army's Mad Scientist initiative, the next war is being planned as one of bio-convergence: "Humans will become part of the network connected through embedded and worn devices."[91]

Like the original cyberneticist and war innovator Norbert Wiener, Whitehead was interested in "the living body as a coordination of high grade actual [i.e., biomaterial] occasions." He thought about experience as subtended by "centers of reaction and control."[92] Though Whitehead leaves it for others to surmise, the relation between the quantitative self and the state—particularly as the state manifests visible and invisible forms of political violence—seems to shadow his discussion of "process and reality" in the book whose title is those very words. Whitehead's subtitle for the book is *An Essay in Cosmology*. And notably, in Whitehead's account, the cosmos is not a superficially peaceable place. It involves a good dose of what the philosopher of science repeatedly refers to as "unrest."[93] This is the same unrest James identifies in relation to Bergson's philosophy of mind when he insists upon a carefully thought-out "return to empiricism."[94] For James, a renewed attention to matter revealed a politically vertiginous, quantum-like property, something he simply called "motion." But it is motion as "an originally turbid sensation," James continues. "We feel that movement is . . . more or less alarming and sickening."[95] A theory of cognition as matter in motion is "alarming" for James because what appear to be inanimate and material processes are fully enmeshed with life and death. In a sense, there is in James a different recognition of Kant's graveyard of the human race than what Kant himself offers. Here the living and the dead recombine: zombies, not graveyards, are the prefiguration of posthuman war. James writes, "I think the return [to empiricism] only proves an immortal truth: What won't stay buried must have some immortal life."[96]

In "The Moral Equivalent of War," James thinks further about what will and will not stay buried, as if the empirical, how we succeed and fail by adopting realist philosophical principles, is connected to the line between war and peace. "Military feelings," he writes, "are too deeply grounded to abdicate their place among our ideals."[97] Referring to a former time where "all the citizens were warriors," and in hoping to find "better substitutes than glory," James draws our attention to what he calls humanity's "double personality." He continues, "It may even be reasonably said that the intensely sharp *preparation* for war by the nations is the *real war,* permanent, unceasing; and that battles are only a sort of public verification of the mastery gained during peace."[98] To James's realist doubling of the human being per se as a concept at war with itself, we can add a third component, one equally operational in the so-called zones of peace. The latest so-called real war becomes more explainable in a philosophical language spanning the idealist divisions between human being and machine, mind and matter, epistemology and military conflict. The tripling of the human personality, or better put, its multiplication in the form of computational renderings of thought, is already marked by realist philosophy. Computation is not just the second element, doubling humanity apropos James; it is every other number thereafter. James laments, "Our ancestors have bred pugnacity into our bone and marrow, and thousands of years of peace won't breed it out of us."[99] We miss the opportunity to understand permanent, unceasing war at full scale if we take his reference to the human body to be an obligatory ethical gesture. Bone and blood, clearly; but in addition, our coevolution with computers allows war to be bred into brains.

At first seeming to echo Kant's transcendental humanism, Whitehead writes in *Process and Reality* that "in the philosophy of organism it is not substance which is permanent, but form" (29). He seems to echo Kant because placing form over substance appears to embrace the categorical imperative. Genus matters more than *matter* because it contains and unifies the otherwise alarming multiplicity matter always contains. This would contradict certain non-Kantian discoveries about cognition—namely, the ones Damasio and the Orch OR researchers posit where thought is an effect of quantum vibrations. But Whitehead also says (quoting Locke), "entities 'perpetually perish,'" which, you could add, is why there is no such thing as perpetual peace, and maybe where James finds his concern about humanity as a war species. Whitehead counters Kant's categorical imperative by placing the function of matter over the "form" of mere thought. This is precisely the move the U.S. State Department attorney Koh makes in defining the enemy as relational, instead of categorically univocal. In a chapter from Whitehead's *Process and Reality* called "The Categorical Scheme," Whitehead directly contradicts the idealist basis of "Kant's First Analogy of Experience" (28–29). Contra Kant, he offers an investigation of subject/object relations along Bergson's realist lines. This opens human experience to its relationship with organisms beyond mere human—and even beyond mere animate—ones. Whitehead attempts to focus "not on the individual entities of the kinds, but the collective kinds of the entities. . . . The 'universe' comprising the absolutely initial data for an actual entity." Further, "a multiplicity enters into the process [of making sets] through its individual members. The only statements to be made about multiplicity is to express how its individual members enter into the process of the actual word" (29).

In this last citation, the word "actual" denotes both Bergson's notion of matter and Cantor's emphasis on infinity in the kind of set theory we described in reference to Badiou. It is also consistent with a quantum theory of cognition. Whitehead calls this emphasis on collectivity over genus by a name we now know to be familiar: "multiplicity." This same "multiplicity" is what designates the war-within-peace problem I was referencing by rewriting Bennett's title to read *Violent* rather than *Vibrant Matter*. Note, too, that for Whitehead, "multiplicity has a solely disjunctive relationship to the actual world" (30). The affirmation of multiplicity as disjunction is why neither Whitehead's nor Bergson's realism (nor indeed, Badiou's) presents a so-called *flat* ontology: a straw charge laid at realism's feet where, because everything is everything, no one thing is identifiable at all. The better way to channel this temptation toward philosophical impasse is to say that no one thing is immutably definable because no one thing ever exists as just *one*. The disjunction Whitehead refers to simply means the categories by which we organize and produce knowledge change. Moreover, they change within the context of specific moments of conflict and struggle. They are dependent upon the human being at war with itself. This is how realism reworks Kant's almost morbidly ironic depiction of a war to end the human species, its demise as one form of perpetual peace.

OPTO-ELECTRONICS

If we return yet again to Clausewitz's famous assertion that "war is politics by other means,"[100] and especially if we realize that *means* equals *minds*, then we are squarely within neuroscience as part of the tripling of the human personality and beyond.[101] DeLanda and Harmon allude to realist philosophies of mind as having military applications when they suggest that "the mind-independent [i.e., not subjectively constituted] world of matter and energy share the space of the battlefield."[102] But there is no better-known philosopher of media and war than Paul Virilio. Though the list of his relevant works is too long to review here, his interest in the image, rather than the more commonly noted writing on the politics of speed, has relevance in the context of the brain's wetware as a virtual-reality-rendering computer. Through the digitally mediated image, human cognition is added to the arsenals of war. Thus, Virilio insists, "we must take hold of the riddle of technology and lay it out on the table as the ancient philosophers and scientists put the riddle of Nature out in the open, the two being superimposed."[103] Virilio's addition of military violence to the nature–technology "superimposition" is well known. I will not summarize it here.[104] But we must consider if the distance Virilio goes in solving the technology riddle has ultimately fallen short. In rendering war total, Virilio is both too extreme and not extreme enough in the kind of *totalizing* he offers. On the one hand, he is too extreme. He reduces computational and visual media flatly as manifestations of military violence; on the other hand, Virilio is not extreme enough. By regarding computational technology as only another means of war, he misses a more nuanced (Jamesian) point about war existing within peace.

Notably, in thinking through the nature–technology pairing and its relation to military neuroscience, Virilio starts with a Bergsonian notion of cognition as image-based. This hypothesis is borne out later by the computationally oriented brain research already cited. For Virilio, vision is correctly theorized along cinematic lines as the movement and organization of matter in space and time. We have already cited these virtual dynamics in the form of the quantization of data at play in the movement of electrons flowing through bodies and machines and operating according to dynamics consistent with materialist explanations of reality. Virilio is in line with this theory when he and Lotringer remark in *Pure War,* "Technology is our new nature"; and therefore, "'Epistimo-technical' is a very good word" (28–29). The word is good because human cognition and machines work in common as information systems. As an effect of the larger system achieved by combining the two, virtual reality is the medium within which all the usual binaries—self and other, subject and object, animate and in-animate, life and death—push and pull in relation to each other. But if you follow the mind–media–matter connection all the way to where it leads, you will have to replace the middle word "and" in all those pairings with the word "in," as in the biggest pairing of all: the war-in-peace. To the same degree humans become terrain as theorized by counterinsurgency theory, the mind is no longer to be regarded as a simply human organism: mat-ter lives (or at least vibrates), as we have also already seen. To realize the mind–media–matter relation is to admit the contribution of force in how human beings relate to the world and each other.

A better way to put the force–matter connection, implicit within Vi-rilio's nature/technology overlap, is to say that force is a matter of *matter* itself. It exists as a fundamental property of what Virilio and Lotringer call mass. Because they have the primacy of computation very much in mind in evoking such a politically *and* scientifically loaded term, they quote Napoleon, rather than the more likely figure of Einstein, writing after the French military master: "Force is what separates mass from power" (37). Napoleon is the right choice to think of mass as force, if your topic is how masses are mobilized in war. But if force can be read here in the sense of being equally operative in the laboratory as well as the battlefield, then we can also say that force is linked to knowledge, and knowledge is linked to war. The U.S. Department of State's concept of the functional combat-ant, where the enemy is construed as a target even before having violent intent, reflects upon Virilio's reference to the "information bomb" as a temporal as well as a spatial weapon.[105] The treatment of time as a wea-ponizable resource—something to be mined, stored, and deployed, like all hazardous substances—is not new to war. This is why Napoleon is a good source for media theory. But the degrees to which time may be ex-panded and compressed are new because our tools (and our weapons) are more capable operating this way. Virilio thus speaks of the century of the image as the birth of media modernity and as contemporaneous with the practice of modern war. Together, war and cinema add up to what he calls "an intolerable situation in which industrial optics have run wildly out of control" (29). The intolerance is also why he alludes to—without quite completing—the connection between the technological bases of the "new opto-electronic arsenal" and the reconstitution of the human being as "a cybernetic entity" (40).

The reason I say Virilio alludes to, but does not complete, the connection between opto-electronics and the human-being-as-data-entity is because for him technology is almost always bad. It is always on the other side of the human being, our nemesis, our ultimate undoing. This, I want to suggest, is a mistaken media theory keeping the human being cordoned off from the information platforms we depend upon even as they may destroy us: if information equals war, and war is bad, then media technologies are also bad. But can we be outside media any more than we can be outside brains, as this badly constructed syllogism suggests? No more than we can be outside of war. Virilio's alternative to war as the escape from computational and visual media keeps him from seeing the invisible war going on *within* peace (i.e., the violence *within* civil society, *within* the human being per se, and *within* cognition itself). This, too, shares a certain irony with Kant, because the invisibility of contemporary war is what Virilio's fatalistic take on the camera wants to make clear. His take on war-and-the-image is partial: the *humanist* form of the human being haunts his work like the lost ideal it has been since Kant. "Information highways" create "disturbances" rather than go to new places rife with both peace and war (39); "the unstable image is of a fleeting nature," rather than being an information practice capable of expanding and preserving new realities;[106] through the "image," we have "lost," not gained, "dimensions";[107] and therefore, "virtual reality" means "the disappearance of matter," instead of being a commonly shared media situation where life and death share an expanded material platform.[108]

Virilio's analysis of "lumiocentrism" is dependent on a modern war mechanism, namely the camera, but also on an account of future wars within cyberspace and virtual reality. As he poignantly suggests, these digital wars will present a situation of "morphological irruption" and "iconological disruption."[109] Disruption yes, but light on the brain creates memory too. Neuroscientists tell us a specific memory can be encoded and recoded within a confined ensemble of neurons. By the optogenetic stimulation of certain regions in the hippocampus—lighting up the brain—researchers are able to activate and deactivate mnemonic processes.[110] The mind can also be *tuned* by the application of a "Neural Dust that uses ultrasound waves to power tiny wireless sensors that can be implanted in the body and read its output."[111] The bioactive dust mote contains information from the electrical activity of a given cell, but it can also create new information by sending radio signals directly to the brain. For example, fear can be created, or erased, without any frightening event (save a chemical one) having occurred. By applying molecular electronics to specific points in the brain, molecular switches can be flipped on or off by the application of near-infrared light at the convergence of cognitive science and optical physics. In this way, light can be directed to the brain of mice to make them feel docile. Alternatively, this application of light within the brain can turn them into lethal hunters.[112] The same physics-based subfield of molecular informatics uses three-dimensional atomic structures to compute data in nonbinary ways. Virilio theorizes the optical logic of lumiocentrism by giving full credit to the physics of light but misses the full-scale complexity of optical physics. His theory of the image is theorized within the same pessimistic limits he brings to technology in general. On the one hand, luminosity is conceived as a material agent rupturing time and space as human beings ordinarily experience it. It is exactly this explanation of light offered by quantum theory: light is composed of photons, which are haunted by the shadows you can only understand by looking through optical machines. But on the other hand, Virilio implies we should be able to imagine space and time independent of their mediation. Only with such independence can the human being be conceived as independent of political force.

Virilio posits too flat an opposition between virtual reality and actual reality (his terms), where one reality pushes out the other for the purposes of perpetual peace.[113] "War is cinema, and cinema is war," he writes.[114] And further in *War and Cinema*, "The intensity of automatic weaponry and the new capacities of photographic equipment combine to project a final image of the world, a world in the throes of dematerialization and eventual total disintegration" (73). But the theory of the image, from Bergson and James to Damasio and quantum-based neuroscience, insists on a materially variegated expansion of matter as part cognition. This is opposed to Virilio's disintegration of the world. When he remarks that "totalitarianism is latent in technology," despite a little restraint by the word latent, he offers a one-sided account of the vision machine. We cannot account for war-within-peace from this standpoint, since the two are separate in a way inconsistent with the role media plays in military violence. The "camera's peephole" may lend itself to the "tactical necessities of cartography," and no doubt serves as an "indirect sighting device complementing those attached to weapons of mass destruction" (1). But to insist that "the image" symptomatically "gains sway" over "the object," and "representation" usurps unmediated reality, is to miss the realist lesson of how objects and information are brought forth and intertwined. Possibly worse, such a notion hides rather than reveals the ubiquity of war. The "de-realization of military engagement" (1)—and with it, cinema's "trade in de-materialization" (32)—means technology's first and only goal is the annihilation of the real. But the machines of war and peace reveal a great deal more reality than we would have without the tools bringing it forth. This reality exists in places overlooked by centuries of idealist thought, and it will continue to exist whatever human beings do in the way of preserving and destroying one another.

Thus, contra Virilio, the philosophers, scientists, and engineers who sought an explanation of the mind–matter connection as he did—in the context of war—found a greater reliance on the image insofar as it meant reality's *re*-realization. Matter, they say, exists in functionally influential ways, and exceeds traditional accounts where war is wrongly presumed to exist outside bodies, ideas, and the liberal formation of the state. War exists by invading these areas, in certain invisible locations, where we may miss them if we do not use the right tools. To say that these tools are also weapons means that, contra Kant, we can come to terms with the war-within-peace. If the "eye's function is the function of a weapon" (3), then you may start to worry—unless you concede to being a warrior—if it is worth seeing anything at all. Virilio makes a prophetic point next to this one: brains too are tools working like the camera's eye. He writes, "Weapons are tools of destruction but also perception, just as stimulants that make themselves felt through chemical, neurological processes in the sense organs and central nervous system, affecting human reactions and even the perceptual identification and differentiation of objects" (6).

It is true that Goebbels loved *Gone with the Wind* (8); and U.S. High Command enjoyed the services of John Huston, Anatole Litvak, and, more remarkably, Luis Buñuel, before military violence was reconceived as electron versus electron. But Douglas Kellner is right to suggest that "Virilio has a flawed conception of technology that is excessively one-sided" ("VWT" non-paginated). He writes more dramatically: In "mourning the loss of the object," Virilio is "shrilly techno-phobic and consistently hysterical"

("VWT" non-paginated). By contrast, Kellner remarks that "the emancipatory and democratizing aspects of new computer and media technologies" ("VWT" non paginated) must be emphasized instead. But neither Virilio nor Kant, nor Kellner in this statement, are able to identify war in the capacious sense of existing within "emancipation" and "democracy." Twenty-first-century war reveals a struggle between war and peace taking place not so much within a context of reality gone bad with technology, but under conditions of reality mediated in the multiple ways it always has been, through technical means applicable to war and peace at the same time.

VIRTUALITY AND WAR

Here it is important to evoke Bergson one last time—specifically, his theory of war. In *The Meaning of War: Life and Matter in Conflict,* written at the onset of World War I, Bergson uses the term "systemic barbarism" to explain the relationship between Prussia and Germany.[115] His interest in the theory of knowledge as stated in the auspicious year of 1914, and especially his connection of meaning to matter and conflict, should indicate a kinship between war and the mind–matter–media relation. Bergson writes that the dictator's "evil genius" was to get "the German people to believe in a permanent danger of war." "Instead of dissolving Prussian militarism into her own life," he continues, the German state "reinforced it by militarizing herself" (10). Moreover, Hitler cultivated a sweeping "breath of hatred . . . against a common enemy" (10). For Bergson, this "hatred" was activated within the conditions of "material prosperity" and "boundless ambition," where "rights" are nothing more than "concessions for capitalists" (11). In this way, the German nation "simplified its aims, and reduced them to materialism" (9). With "brute force, [and] with its train of trickery and lies," the genocidal state operated as if "it had within it a hidden virtue, mysterious—nay divine" (12). The origins of the fascist state, which imbued governmental power with a godly aura, ran its course alongside the sanctity of the wealth of nations: "People fight to avoid starvation, *so they say*. . . . But without being exactly in danger of starving to death, people consider that life is not worth living if they cannot have comforts, pleasures, luxury" (289, emphasis in original).[116]

Bergson's emphasis of the phrase "so they say" homes in on the war-within-peace existing in Germany at the time. It also draws attention to the virtual war and the German people's belief in the ubiquity of military violence: the need to publicize the strength of the state in the form of repeated mass rituals, the persecution of the internal Jewish foe, and the obedience to one dictatorial master, all before mobilizing for actual combat on a world scale. But throughout his work, Bergson's notion of virtual reality is different from Virilio's war-as-cinema theory. In contrast to the strict denunciation of a virtuality as object-loss in Virilio, Bergson objects to the German state operating as if "force and right were the same thing" (12). This is consistent with his objection to capitalist hypocrisies—attendant to "the great expansion of industry"—where desire is raised over need.[117] But there is a further-reaching point contrasting Bergson's materialism with the idealism lingering in Virilio's notion of the information bomb. Bergson's challenge to Hegel and Kant is consistent with his theory adjoining knowledge and matter, but again, he does not use the term "virtual," as Virilio does, merely as an application of war. His notion of virtual reality, extending to his war writing, is consistent with the visibility and invisibility of war in human history, as well with how wars exist within peace.

Bergson's reference to virtual reality and war is related to his enigmatic use of the word "soul." He writes, for example, that it is the "original structure of the human soul that modern man . . . be closely united." But this is not Kantian idealism, since he also says that in seeking such "unity," there is "between group and group a virtual hostility."[118] The logic is unusual here, and conjures a notion of war mingling within the peace of the commonwealth. As Foucault suggests, Hobbes's famous recognition of "the war of every man against every man refers to the most general war of all, and it goes on at all times and in every direction."[119] The emphases on the phrases "all times" and "in every direction" joins Hobbes to Bergson insofar as both are interested in virtual war. As in Hobbes's fear of the multitude's turn to force within the people, Bergson's use of virtual prohibits flat opposition. The word "virtual" does not signify a binary relation between friend and foe, subject and object, war and peace, or any other idealist dyad. To the extent that virtual hostility gains political inflection, enemies are neither essentially nor eternally opposed. There is always the potential for slippage, crossover, a sudden change between sides, and the unexpected shifting of boundaries. Citizen and combatant are virtually proximate. Accordingly, they can be hailed into existence as one or the other, or both at the same time, in an entirely legal, though potentially disastrous, way.

This ties virtuality to Bergson's account of capitalism's enabling role in World War I and extends his theory of human cognition. The capitalist materialism he wishes to indict by name is prone to violence not just because it is based on science but, more specifically, because it is based on the wrong kind of scientific reason: science in idealism's disguise. Bergson elaborates on brute force as a perverse use of tools to foster luxury for a handful of sovereign elite. Further, he wishes to correct a misunderstanding about the relationship between life and matter, which is what leads to fatal consequences as produced by the commodity form. Science in this instance is the opposite of capital, and not luxury's military enforcer. Next to Bergson's indictment of "luxury," and as evidence of the invisible and visible wars Hitler cultivated in the name of prosperity, Bergson says in *The Meaning of War,* Germany went wrong when "it translated her amazement [at wealth] into an idea" (12).

For Bergson, consumerism in political form supported the predatory nation's territorial conquests and was mistakenly presumed to reflect the earth's geography as compatible with German racial purity. He continues: "See how, at the call of this idea, a thousand thoughts, as if awakened from slumber, and shaking off the dust of libraries, came rushing in from every side" (12). The enigma of so-called Aryan identity under Hitler lies less in its militarized appeal to racial superiority than in its ability to draw in— from every side—the massively complex reality of human difference, and give it unified form. In a particularly forceful critique of idealism, Bergson writes that this is how "Germany . . . invokes Hegel as a witness"; and "would have declared herself faithful to Kant" (12–13). Bergson repudiates "ideas becoming official doctrine." Idealism, he suggests, was doomed to "translate into ideas what was . . . [in reality] insatiable ambition and will perverted by pride" (12). Bergson's indictment of idealism in its most horrific (fascist) form makes it hard to reduce his antiwar philosophy to the spiritual influences of St. Augustine.[120] There is a role in Bergson for the Augustinian caveat in *The City of God* against "the lust to lord over one another."[121] Lording is part of the problem for him, not the solution; and it is an epistemic as well as a material problem, with capitalism as idealism gone wrong, and with war as the result of idealist notions of the citizen as being of one race.

Bergson's realist conception of national division shares a theory of category close to Cantor's set theory and is a theoretical forebearer of Badiou's interest in "numeracy." All three theorists help explain the notion of the functional versus the formal combatant in contemporary war jurisprudence, because for them any presumed unity contains elements that lead in the direction of its potential disruption. All categories, epistemic as well as ontological, are virtual because they have other categories hidden within them. "Nature," Bergson writes in *Two Sources of Morality and Religion*, "could have found no surer way of making every foreigner a virtual enemy" (247). But before too much stress is placed on a bad definition of nature as unchanging, outside of media, or making people have innate behaviors, Bergson calls on virtuality to reveal how nature, too, makes things (as in making human beings) in virtual ways. With clear optimism, he writes, "Anyone familiar with the language and literature of a people cannot wholly be its enemy" (247). The key word in this sentence is neither "language" nor "literature"—for which we should simply say "media"—nor is it "people," nor too simply, the "enemy." These are important terms for reasons already made clear in Bergson's difference from Virilio and his theory of cognition as a phenomenon of mind, media, and matter. To read him accurately, we must hold fast to the supremacy of quantization over the categorical imperative. The most important term behind Bergson's optimism is "wholly." There are always more parts left over in group-to-group, friend-to-enemy, relations. Because those parts exceed the opposition falsely presumed to be whole, any hope for peace in the war-in-peace relation must take place first in a virtual way.

In Bergson's war writing, the war-in-peace relation exists within the supposedly peaceable relationships of the market economy, which perpetuate war both seen and unseen. Thus, Bergson remarks, "the origin of war is ownership, individual or collective" (254); and "since humanity is predestined to ownership by its very structure, war is natural" (254). The phrase "by its very structure" qualifies the more striking stipulation: war is natural." The decisive link between war and luxury in the commercial socius is this: "If the self-reproduction of man is not rationalized, as his labour is beginning to be, we shall have war" (250). About the social contract, Bergson writes: "Man outwits nature when he extends social solidarity to the brotherhood of man; but he is deceiving her too" (44). The word "deceiving" designates the way in which the concept of the human being per se has really never been external to the generation of wealth vis-à-vis military violence. "Because idealized self-reproduction is based in buying and selling the work of the many, with its decisive relationship between reproduction and women's work, and because the capitalist economy leads to excess in the hands of the few, Bergson is taking issue with a multi-layered system of false equality. In economic terms, false equality means exchange value over use value, what you can *buy* over what you can *do*; what you *want* over what you *need*; what you *have* over who you *are*. Bergson objects to these false equalities at the level of individuals in relation to each other. The wealthy do not have the same interests as the poor, and in effect wage a war within peace against them under the false heading of a common humanity. Bergson rejects a version of modernity tethered too closely to capitalist ideals. But he does not retreat from either materiality or reason into the kind of spiritualist moral critiques other readers have assigned to him.[122] Instead, Bergson offers a realist theory of civil society under the market system—the war-in-peace first discovered by Hobbes—that is consentient with his materialist philosophy of thought.

Bergson's initial lecture, the basis for his book *The Meaning of War*, was more interestingly titled "Life and Matter at War." Property is described here as the canalization of material abundance leading to the accumulation of wealth in grossly unequal ways. But the more general problem of getting multiplicity wrong is also addressed in ways adjoining Bergson's writing on war with his interest in the mind–media–matter relation. For Bergson, Hobbes's war-in-peace exists within the widely applicable utilities of what he calls "invention," but also, variously, "engine," "mechanical super-imposition," "automatism," or simply, the "well-appointed machine" (9–11). In the form of the German state, the technological applications of war happen at the level of assembling political relationships. The dynamics are the same as in his theory of the human brain.

In the case of the German Empire "on the morrow of the war of 1870" (11), formerly independent countries are redivided and reorganized, subsumed within new geographical entities, that expand the boundaries of unity even while German nationalism heads toward collapse. State orthodoxy proceeds to organize life and matter at both the micro- and the macroscopic scales. But the organizing faculties of government are insufficiently capable of dealing with reality at a level of complexity that actually exists. Here Bergson discovers a certain technology-of-state predating Foucault's more famous governmentality lectures. Concerning the "task of her [Germany's] organic self-development," he writes: "There was within her, or rather, by her side, a people with whom every process tended to take mechanical form." "Artificiality," Bergson continues, "marked the creation of Prussia." Moreover, Prussia's relation to the German empire was "formed by clumsily sewing the two together." In the rest of the passage, Bergson's key word is "machine": the administration of the German state is a "well-appointed machine." It is as "mechanical as the army" (9), and evokes the idea of Prussia as "a vision of rudeness, rigidity, and automism" (9). Germany had to "choose"—and the state chose badly—"between a rigid and ready-made system of unification, mechanically super-imposed from without, and the unity which comes from within the natural effect of life" (9).

What about the biological unity called life and the apparent dichotomy here between the so-called natural form of organization and what looks like its opposite, the mechanical one? It would be tempting to see natural life in this instance in the idealist sense, even though in the passage above Bergson is characteristically down on the "*idea* of Prussia." He calls this a "clumsily sewn" unity, a national identity riddled with seams. But Bergson is not taking issue with machines at large, no more than he is taking issue, on behalf of idealism, with science; nor is he dismissing matter by critiquing luxury as a capitalist ideal. As he states clearly, machines can go wrong. But this does not suggest that life is best lived without them. Bergson is concerned about military violence as immanent to social existence when "every invention is bent from its intended use, and converted into an engine of war" (11). But the bent machine as the war machine leaves room for distinguishing better from worse forms of bending. Bergson rejects the misapplication of technology at the political level when he indicts predetermined patterns of geological sewing on the order of German territorial expansion. The *matter* of a nation in this sense (its political geography) exceeds the unity of the nationalist *idea*. Where the German nation sought "unity in diversity," it failed to "create a closer union of confederate states" (9). It failed by imposing draconian forms of nationalist identity on a people whose way of self-reproducing did not in fact follow such rigid logic. For Badiou's numeracy to find its proper way, nationalism must necessarily collapse. Bergson says simply: "Time is needed for that" (9). Economically speaking, industrial machines in the right hands are capable of producing a more equally distributed quantity of material goods and can produce the leisure time away from work needed to enjoy them. To Bergson's anti-idealist eye, those machines have failed to deliver things as a result of industrial capitalism amplified by nationalist ideals. But this does not mean the machine-as-such is reductively a war machine. Bergson's theory of technology posits the machine as working on behalf of war *and* peace.

Bergson does not exhibit the kind of technophobia Kellner finds in Virilio, nor is his theory of reality opposed to mediated existence. On the contrary, technology for him is not so far away from the mind; and, as we have already seen, cognition depends upon the movement of matter. To think that human beings live and die without tools is impossible for Bergson because the mind itself is a kind of tool-conceiving technology on its own terms. Ultimately, Bergson counters the bending of machines in the direction of Nazism not by rejecting technology but by positing a better mind–media–matter relation. In contrast to Virilio, he remarks in *Two Sources of Morality and Religion,* "Man must set out simplifying his existence with as much frenzy as it devoted to complicating it" (226). Bad versions of complication include pursuing the kind of war-within-peace implicit to wealth inequality. But this kind of war is "not inherent to the machine" (266). Bergson questions "ascetic . . . mysticism" because the real complexities of matter are better than the idealistic ones driving the inequalities of wealth: "Man will only rise above earthly things if a powerful equipment supplies him with the requisite fulcrum" (267).

Bergson despises "luxury for the few, rather than liberation for all" (267), as we have noted, and he attaches luxury to a war-within-peace. His opposition to debased materialism insists upon more powerful machines, not their dismissal. But they must be machines directed toward a new and better materialist order. He remarks, humanity must "use matter if [it] wants to get away from matter" (267). The paradox behind the enigmatic phrase "get away from matter" can be explained by how Bergson wants "machinery to find its true vocation" (268). But the true vocation of media goes in at least two directions for Bergson. One of those directions has to do with scaling up reality to conceive of better futures; the other has to do with the related issues of categorical change. Bergson's pitch for "machinery to find its true vocation" is, in the first instance, a matter of "proportion." "We must add," he says, "that the body now calls for a bigger soul, and the mechanism [by which we go big] should mean mysticism" (268). But how can we reconcile Bergson's rejection of the "mystical ascetic," his affirmation of machines, and his critique of luxury, with this rather bizarre-sounding embrace of mysticism? What can his adjoining of mechanics and mysticism mean? How might it help us use Bergson's insights on the fascist war-within-peace to understand the same phenomenon (though expanded) in posthuman war?

Bergson puzzles us further: the "process of mechanization is more mystical than we might imagine"; and further still, "the mystical summons up the mechanical" (268), and thus produces a "true mysticism" (269). To understand this concept of mechanical mysticism, we must keep in mind Bergson's firm opposition to "luxury for the few" and his support for the "liberation of all" (267). The term "true" is used here in a highly nuanced manner to "encourage a very special 'will-to-power' that will make of it [humanity] a new species, or rather, deliver it from the necessity of being a species; for every species means a collective halt, and complete existence is mobility in individuality" (269). In this distilled passage, the idealist implications of mysticism are ruled out by the way Bergson challenges the idea of the human species with the term "more." Alongside "more," Bergson's other keywords—"collectivity" and "mobility"—foreshadow the technological transformation of the human being and give us theoretical insight on a new form of war within the human body. Bergson posits an expanded version of humanity capable of self-reproduction beyond the demands of capitalist accumulation. The species of the human being per se moves toward liberation by achieving greater forms of collectivity, where the term "collectivity" includes both humanity and matter. "Mechanization as it should be," Bergson continues, "means the workman's tool is the continuation of his arm, where the tool-equipment of humanity is . . . the continuation of his body" (267). Here Bergson optimistically affirms a "very special 'will-to-power,'" where the technical and human evolutionary processes work cooperatively and in tandem. But this does not mean he has abandoned the war-in-peace. To the contrary, Bergson offers an invitation to struggle within war's ambivalent future, the realist's ability to express as much hope as fear.

WHITE MATTER

By placing strong emphasis on technology in the way that Bergson insists, we can better identify the stakes of his focus on the war-within-peace as an alteration in the human species. In this sense, he was working in a philosophical way within the mathematical field later established by Norbert Weiner: cybernetics. Weiner, whom we have already introduced as a foundational thinker of the mind–media–matter relation, was interested in a specific element of brain activity. Here he is interested not simply in grey matter but also in the more important component for understanding brain circuitry: white matter. The discovery of white matter is particularly applicable to war neuroscience. It is also a convenient way to sum up how far we have come on posthuman war. From the fluid demography of a postwhite America imagined by U.S. census politics to the invention of the white Afghan by the HTS program, whiteness has taken on demographic as well as anthropological significance. We have seen whiteness operationalized for military violence in both the ontological and cultural domains. But this begs a third question about posthuman whiteness: What happens to whiteness at the level of human cognition as racial identity shifts into multiply networked brain dynamics, a kind of cybernetics 2.0?

Wiener's focus on white matter in the brain is appropriate for a computer scientist. Whiteness here refers to the myelin substance coating cells within the nervous system. This coating directs the electronic flows to their neuronal targets and is the brain's most important material for it to communicate among its own regions, with other human beings, and with machines. The mass of white matter in the brain outnumbers neurons by ten to one, and is regarded not simply as a locus of data exchange but more important as a kind of information mediator. Wiener writes in *God and Golem*: "If the white matter of the brain is considered equivalent to the wiring of a computer circuit, and if we take each neuron as the functional equivalent of a transistor, the computer equivalent to a brain should occupy a sphere of something like thirty feet in diameter" (72). "Actually," he continues, "it would be impossible to construct a computer with anything like the relative closeness of the texture of the brain" (72). Weiner seems to be arguing for a separation of, rather than a proximity between, mind, matter, and media, contradicting his theory of information as a dynamic of command and control, and undermining the very premise of cybernetics. But the problem here is the technical limits of a brain-machine interface in the limited ways it existed in his day (1964). Wiener's broader point is more important than this. White matter is referred to by his neuroscientific heirs in exactly his terms as wires, and as the information superhighway of the brain. The breakthrough of cybernetics was to identify compatibility between the way information is processed by human beings and how data are processed by other computationally based entities. Both operate like Bergson's virtual-reality machines.

Wiener worries about the "enormous operational size" necessary to see media and minds "actually" overlap (72). While he did not surmise the coming war neuroscience, his interest in white matter as biological media foreshadows DARPA's interest in the human brain as part of a new military arsenal. A Defense Science Board task force document adds to an earlier—and waning—interest in the mere cultural work of Human Terrain System, literalizing the keyword "terrain" to mean specific territories and assemblages within the soldier's body. War neuroscience designates a geologically inflected focus on "human dynamics" in the electrochemical sense.[123] "Attitudes, influence networks, and the effects of strategic communication" remain of interest, as with the cultural understanding of "terrain." But networks are further expanded to include "bioinformatics."[124] "Open architecture state-of-the-art platforms for data, model, and tool integration" allow the contemporary war strategist to integrate subjective phenomena with biomechanical processes: "sentiment, intention, deception" become their own areas of operation and may be treated as territorial space. Accordingly, "geospatial-dynamic network analysis combines with neuro-cognitive models" as overlapping "network arrangements."[125] In keeping with the geological inflection, neural dust can be sprinkled on white matter as a conduit for sending and receiving transcranial radio signals. These microscopic transceivers placed within the human cortex are called "interrogators." Neuroscience and military activity are converging, and this convergence brings new meaning to terms like networks, populations, data signals, firings, information superhighways, and so on, in the nonmetaphorical sense. Human cognition is cultivated at the level of the synapse as an instrument of posthuman war.

Scientific explanations of white matter use the phrases "computer network" and "the subway of the brain."[126] In the same way that Wiener suggested, it is both an actual—as in, numerically measurable—substance, as well as a "fast . . . transport system . . . connecting different regions of gray matter in the cerebrum to one another."[127] The study of white matter as a means for moving war matériel likens the brain to logistical military problems. Its "tracts" (read: tracks) are physical, not figurative. But "white matter is made visible by a new kind of image technique."[128] Working in the terms Damasio describes where the brain is a generator and coordinator of images and keeping in mind the virtual nature of reality in Bergson's mind–matter–media relation, the term "image technique" designates a materially based form of information flow. A more precise way to explain the relationship between white and grey matter is "projection versus computation," according to neuroscience terminology.[129] The literal quality of these terms cannot be stressed enough. The kind of projection going on in the brain is best characterized as a virtual process, no less real for being based in pictures, because the images being projected depend upon the organization of matter in space and time.

One of DARPA's many neuroscientific projects is particularly indicative of the significance of virtual reality as a concept for integrating human and machine intelligence. The goal here is to project visual stimuli directly into the brain and receive different projections from it. This $21.6 million experiment is designed to communicate directly with the image-producing part of the cerebral cortex by using a miniaturized "light field microscope."[130] The term "light field" means that this device can see into the brain and do things with the brain as a part of the seeing. The light field microscope is thus programmed to read data from a million individual neurons simultaneously, using its own projection of light to stimulate one thousand of them with single-cell accuracy. This affords two kinds of virtual-reality production: the first kind allows access to new parts of the human body, and the second, more spectacular, form allows a solider to see remotely and without opening her eyes. The same principle of technology is being used in a revolutionary form of prosthetic haptics. This mechanical extension of the arm uses the same kind of electronic stimulation algorithms to produce the physical feeling of a given object without touching it with your own hand.[131] Defense industry biotechnology connects mind and machine in material ways while transcending the distinction between physical reality and mere representation.

Indeed, white matter was once itself invisible to scientists. For a long time, it was neglected as passive tissue, rather than identified as a cognitive and behavioral organ. But innovations in brain-imaging technology, known as Diffusion Tension Imaging (DTI), use a host of fast-acting algorithms reaching beyond the magnetic resonance imaging devices (MRIs) available in most hospitals. New machines allow scientists to capture frequencies passing through computational tensors, which are water-diffused signals whose frequencies are too low, and too asymmetrical, for MRIs to read. As DTI more effectively reveals, white matter is composed of bundles of axons, rather than the more famous neurons and synapses of grey matter. The bundles themselves are "composed of millions of communication cables, each containing a long, individual wire, coated with a white, fatty substance called myelin" (56). Although white matter is referred to as a medium, the color white comes from what insulates the electronic membranes of the axon wires, and either reduces or extends their electronic length for both AC and DC signals. Both the length of the wire and the efficiency of the insulation determine a given current's appropriate speed. If the wire in question is not white enough—that is, if whiteness occurs irregularly in certain brain regions or is malformed along the axon wire— high-level cognition is adversely affected (examples range from autism and schizophrenia to compulsive lying).

Whiteness in a neurological sense performs a supremely integrative function, but on a scale and at a velocity that is not conceivable without one network (the brain's) plugging into another (the machine's). In grey matter, "memories are stored." But in white matter, "electric signals . . . jump swiftly down the axon, node to node" (56). The signals here move at super-high velocities through axons wrapped up to 150 times between every node with living electrical tape called myelin. "This irregular pattern illuminates white matter, exposing the major highways of information that flow among brain regions" (56). Without the essential white myelin substance, data signals in the brain leak and dissipate, lessening the velocity necessary for current to move effectively between white and grey matter.

We should say that whiteness, in the myelinated sense, is the literal application of a color dynamic already happening in the demographic and anthropological applications of posthuman war. In the case of U.S. census politics, of the so-called white Afghan, and in myelinated whiteness, human identity is no longer a subject position. It is instead a fluid substance—

literal fluidity, in the case of "white matter." Something has happened not only to the technologies of war but to the nature of representation. Representation may now be regarded as a technological pathway pushing identity in the direction of numbers, terrain into culture, mind into matter, and ultimately, war into peace. In the expansion of irregular warfare along these lines, military operations include not just a mutation of the human species but also the weaponization of entities, biotic and nonbiotic, seen and unseen. Current war literature calls this expansion of the battlefield by its appropriate epistemological name: simply "human dynamics knowledge" (*UHD* 12).

But this dynamic is curious insofar as the "human" named here is supplanted by "knowledge" as a substance of both body and mind—network-centric information systems, human and machinic without functional distinction. Apropos the preservation of the species, we could say, on the one hand, white matter is a fundamentally human property. For example, 42 percent of the human neocortex is white matter, while in the hedgehog it is only 13 percent ("NCP" 622); and further, "along the invertebrate evolutionary line, the use of bare axons imposes a natural insurmountable limit to increasing the processing capacity of the nervous system."[132] On the other hand, what we are talking about is a form of interconnectedness that not only changes the concept of the human being per se but also removes the humanity–technology division. In this sense, the "enhanced granularity of strategic, operational, and human dynamics . . . becomes an integral part of the war planning process."[133]

The network-centric dynamic we have seen in war demography and war anthropology applies equally to the *terrain*-ing of the human brain. This is most apparent in the form of "bio-electric repetition" techniques (59). As in posthuman war's cultural applications, computational technologies have opened unparalleled opportunities to rethink time and space. But the space-time manipulations at work in the military brain-machine interface concern the electrochemistry of cognition. This is not simply knowledge about the battlefield but the production of knowledge as part of the battle. The awkwardly named Brainternet program is one initiative seeking to translate the several billion kinds of neurological impulses visible on traditional EEG readers into a computer code of zeros and ones.[134] The obstacle is not how to translate thought into data. Rather, it is how to move between the bioelectrical impulses of the brain and artificial kinds of computational systems. Here, again, the scientific discourse around white matter is useful. Whiteness as such coordinates information flows from relatively great distances in the brain, where the electrical current traveling along myelinated wires "arrives simultaneously at one place at a certain time" (59). But the data inputs coordinated by white matter do not just come from great distances in the context of what is—in the context of the world at large—the small distance of the human skull. Rather, the distances in the world at large designate farther distances still, more pathways for white matter to transmit outside the skull to places as far away as the World Wide Web allows.

The material part of whiteness in the myelinated sense should indicate something in common between thought, computation, and the laws of the physics. "White matter can influence cognitive ability," and if damaged is responsible for several forms of neuro-impairment, like multiple sclerosis, cerebral palsy, and Alexander disease (58). It is also key to understanding neural plasticity, the way in which the world of things can influence the brain's "myelination emodeling" capacities.[135] But a conceptual reevaluation of the computational basis of cognition introduces the possibility for discovering what one leading neuroscientist calls a "universal language."[136] The proposals to map the brain at the scale needed to compute outward from the human mind to other kinds of entities requires extremely large amounts of data. For example, the roughly one hundred trillion synaptic connections in the human brain produce on the order of a zettabyte of data (about sextillion bytes). At the point where the term "language" no longer refers to just words, the literal sense of the term "universal" becomes essentially numerical.

More to the point of identifying the nonhuman aspects of human understanding, individual experiences with the world at large can also "influence myelin formation" (58). Information—writ here as electric impulses—must be transmitted at high speeds within the brain and must register in neighboring brain regions simultaneously. For example, myelinated axons move along the neural network in about 30 milliseconds to travel from one cerebral hemisphere to the other. By contrast, an axon wire stripped of the essential ingredient of biochemical whiteness reduces transmission speeds to 150 to 300 milliseconds (59). Velocity is a key factor in calibrating human cognition, second only to the placement of the information being moved around within the brain. "Quantification of connection strength can be challenging," writes one researcher, "but when achieved, it can be considered as providing the gold standard in measurements of large-scale connectivity."[137] The work of white matter is to make sure that the brain's bioelectric repeaters move matter around on a precise schedule. This not only gives it an essential role in cognition, it also presents a logistical situation in parallel with other forms of logistics in the military sense of the term. White matter plays a role in the creation of new categories, extending Bergson's challenge to the human species and giving it scientific weight. White matter has not only a quantifiably determined logistical role but also one of territorial management: "It integrates separated regions of the cerebral cortex" (61). Indeed, the majority of white matter is "inter-hemispheric."[138] "Spatially organized, its inputs and outputs for neighboring cortical regions tend to occupy neighboring pieces of white matter at the same time."[139] In this border-crossing kind of way, the "occupation of neighboring" regions requires us to rethink time and space, as well as category. War's interest in expanding the term "network" to include brain-machine interface is clear on this score: "These tools are mature," the *Report of the Defense Science Board Task Force* reads, "ready to be integrated with other technologies, and expanded, particularly for use in the areas of spatiotemporal reasoning and individual neuro-cognitive assessment" (51).

Neuroscience is not held as separable from cultural studies in the military context, nor "culture" from "demographic attitude" (52), which "combines cultural geography, military geography, and human terrain as a system of subsets of the entire dynamic—changing through time and across nation-state boundaries" (73). These systems share attributes allowing the concept of terrain to function in both material and immaterial ways. Subjects and objects are seen here as existing in common by way of "high fidelity data" (52). While the HTS program uses traditional practices of ethnography to record and manipulate cultural narrative for purposes of counterinsurgency, war neuroscience "detects neural responses underlying empathy induced by stories."[140] Initiated in 2011 and still operational in 2015, the Narrative Networks (N2) program was "created to develop a quantitative approach to the analysis of narratives and their influence on human cognition and behavior."[141] Here again, cognition and media interface within commonly constituted operating systems.

The apparatus of the census enters here, too, but at the cellular level, as a practice of statecraft applied within biophysical science. Researchers working within the U.S. government–funded BRAIN initiative have followed a net-centric method of "non-linear" data analysis to map the human brain. The $500 billion race to map the brain is a significant form of demographic work because the brain's molecular structure is organized the same way as other kinds of populations (*BI* 6).[142] The goal of producing the first comprehensive "cell census network" promises "finer categories of enumeration" than the racial categories of population measurement occurring within the traditional count of bodies and races.[143] An important milestone in the progress of mapping the brain has been the ability to simply name cells and accurately calculate circuits. This is put notably as a mathematical mapping problem, requiring the scaling up of data processing offered by the quantitative sciences to create the first complete brain atlas. In the frank language of neuroscience, producing a brain census provides new access to a more expansive human "parts list." In turn, the expanded parts list keeps track of not only additional attributes for defining humanity but also new categories of being. Diversity discourse at the neurological scale turns against older demographic topographies of race, gender, and ethnicity while dovetailing with counterinsurgency theory's weaponization of identity and culture. For example, only 6 percent of brains in a recent mapping experiment show "substantial variability" corresponding to traditional gender norms.[144] Instead, brain patterns "reveal the inherent intersection of social categories."[145] The way of conceiving social and physical structures here borrows, perhaps unwittingly, from quantum theory. They are "fundamentally entangled . . . and recurrently pass activation back and forth" (795).

The advent of neurodiversity designates the "non-neurotypical" individual as "the latest challenge to accepted ideas of human normality."[146] The U.S. government's presidential study on neuroethics—titled *Gray Matters* for the double entendre in the word "gray"—affirms "the difficulty in defining 'normal'" as a starting point to moving beyond such an outdated term.[147] The point of governmentally encouraged neural diversity is twofold, and is consistent with the expansion of military violence: First,

"normality" is no longer the prevailing way in which the state operates in relation to civilian populations: biopower becomes bioinformatics, becomes
cognition as an instrument of war. Second, nonnormative theories about
human beings do not in and of themselves prevent the de-civilianization
of civil society. What nonnormative theories do is introduce the relations
of force within areas of knowledge and human collectivity that used to
be considered forbidden to military violence. Among the many promises
of war neuroscience is to present a more inclusive population of soldiers
where the logic of inclusion reaches beyond race and gender. Against previous calculations of human measurement, neurodiversity promises a way
of categorizing "individual variability in neural-based traits and tendencies."[148] This "allows us [recruiters and commanders] to break away from
unscientific stereotypes about gender, race, and other previously accepted
ways of categorizing individuals."[149]

The kind of census work being done at the bioinformational level, like
the fluid demography we have seen in census 2000 and the war anthropology applied by HTS operators, involves quantities of information at increasing capacities of scale. But the scale is not merely adding a multiracial
category to the official five races and ethnicities, or introducing the curious
concept of the white Afghan. Quantification exists within war neuroscience
at the level of "petabytes comprising thousands of millions of parts" (14,
32). The discipline founded by this level of computational work is called
neuron ontology, the study of existential agency with an object-oriented
spin. In precisely the language of the counterinsurgency theory, the brain
cell census turns the "*interior* [qualitative state of mind] into the terrain
[of quantifiable neurological matter]" (9, emphasis added). Not only does
"neuron ontology" present an expanded way to construct identity, it also
presents a "new social situation" (10). National security researchers ponder the use of fMRI technology to identify possible insurgents. This raises
the kind of "unanswered legal questions under at least the First, Fourth,
Fifth, Sixth, and Eighth Amendments to the US Constitution," relating directly to the invention of the functional combatant.[150] In its neuronal turn,
DARPA becomes a key coparticipant in the U.S. government's already
well-funded BRAIN initiative. The agency's "*only* charter is radical innovation" (emphasis in original), and for DARPA, neuroscience represents "the
greatest scientific and philosophical challenge ever taken."[151]

DARPA's SyNAPSE program is exemplary of what the agency means by the term "radical." Rather than beginning with biological neural systems and then linking them to machines, this program attempts to make an entirely artificial brain out of computer parts mimicking the expanded parts list of the neurological census. The goal of this project consists of putting together 1.6 billion neurons and 8.87 trillion synapses, which matches the scale of a cat cortex and constitutes 4.5 percent of a human cortex. Information design at this large scale requires a highly sophisticated mapping tool called BrainCam: "a framework that records the firing of all neurons and converts them to a movie for convenient visualization—similar in concept to an EEG trace." The BrainCam processes thought in terms of data pictures: "simulations . . . that are reproduced as alpha waves (8 to 13Hz) and gamma waves (>30Hz) often seen in the mammalian cortex."[152] Such an ability to manufacture neuromechanical infrastructure—specifically the visual cortex—such as that produced by SyNAPSE, portends a range of biosynthetic intelligence available to military markets. Microbiomes that glow (there are 100 trillion to work with), self-assembling biologically based materials, manufactured skin with a full sensory capacity, and more: the new discipline of "biomimetics promises to alternate inorganic/organic nano-layers by introducing suitable concentration gradients in appropriate reactant mixtures."[153] As the terms "nano" and "concentration gradients" suggest, what enables these breakthroughs in the dis- and reintegration of the human being per se is a new density of parts calculated at new levels of scale.

In the language of the HTS program, neurons, too, are targeted as populations in the development of war neuroscience. The DOD-funded program Targeted Neuroplasticity Training uses algorithms to "stimulate certain peripheral nerves, easily and painlessly through the skin, thereby activating regions of the brain involved with learning."[154] By stimulating the brain in this way, researchers claim to boost the memory recall of pictures by up to 140 percent.[155] The program's acronym—TNT—could not be more to the point. It is not simply how the war fighter's brain reacts to explosions on the battlefield. The kind of logistical control going on at TNT's neural scale focuses on peripheral nerve stimulation to speed up learning processes by boosting the release of such chemicals as acetylcholine, dopamine, serotonin, and norepinephrine. The widespread cases of traumatic brain injuries caused by explosive blast waves are treated by the military medical centers that specialize in neurotrauma. Though the language of treating the effect of blast waves on brain waves is based as much in physics as in traditional brain medicine, targeting specific cells in the brain in the TNT program means fighting an offensive neuroscientific war. In a matter of speaking, the brain is also a mechanism—a biomechanism—designed to manage controlled firings. Those firings are informational events, and vice versa. The difference between offensive firings on the order of TNT and the defensive ones of treating blast waves is simply one of scale, and not of kind. Firings in both the TNT and IED senses of the term occur within an expanded definition of the battlefield inclusive of the human body. Here the battlefield is very small, anatomically speaking, even smaller than the smallness of the neuron and the synapse. Once the body's molecular substances become operationalized, DARPA can access cognition as a series of microscopic explosion events. These explosions are concurrent with, can preempt, or can be interrupted by the ones that happen when a roadside bomb ignites. The key to integrating the human brain with weapons systems is not to cordon off the brain from its firings but simply to minimize the improvisational part of the brain as an explosive device.

"Biology" in DARPA's own terms is an "application,"[156] which has resonance with both the media being used to map the brain and the brain itself as media. "Nodal positions" are monitored on the order of insurgent elements, charted as "terrain" or "signal systems," which can be manipulated in the form of short-term memory downloads the soldier may or may not know she has.[157] Memory becomes a military frontier in the same way that counterinsurgency theory turns culture into "terrain." Self- and community awareness in the form of what a soldier or insurgent is willing to recall is sidelined by accessing memory as math. Through the application of what war neuroscience calls Cognitive Augmentation, "the target of memory enhancement is not long-term memory . . . but rather working memory, which encompasses processes used for both storage and manipulation."[158] What is enabling the prospect of increasing working memory in the soldier is to see it on screen as an "electrophysiological event, where changes in the visual field held by an unconscious mind can be turned into information processing."[159] You may have seen where the IED, or sniper, or target was before you shot or were shot. But you may not be aware of this

information. By letting the machine remember for you, your consciousness is no longer needed.

The full military significance of what N. Katherine Hayles calls "cognitive nonconsciousness" enters here. Hayles uses a shorter term, the unthought, to "refer to recent discoveries in neuroscience confirming the
existence of nonconscious cognitive processes inaccessible to conscious
introspection."[160] While the "unthought" may be inaccessible to the subject who thinks, this does not mean the information is inaccessible to the
machines thinking alongside her. DARPA's Biological Technologies Office
is looking toward next-generation electrode-embedded headcaps to apply
high-definition transcranial stimulation tagging memories during learning, and reactivating them during sleep, to improve war-fighting capabilities.[161] The process is known as STAMPing, or "spatio-temporal amplitude-
modulated pattern" making. But there are two patterns being stamped out
by this modulation, the one in the phenomena being (artificially) remem-
bered, and the other in the phenomena being stamped into place within the
tissue of the brain. This is why military researchers are able to use human
brain waves to teach robots how to shoot.[162] Brain stamping works in both
directions, and as such, is being applied to intelligence practices ranging
from foreign language acquisition to increased battlespace awareness.

The enormous volume of imagery generated by geospatial intelligence
overwhelms the human visual system. However, "the development of a
neurologically-based imager triage system" moves military research "toward an operational capability where brain assisted frameworks are integrated into standard satellite-based imagery analysis platforms."[163] Brain
signals can be connected to target acquisition because DARPA's devices
in turn "target memory restoration in the human brain." According to one
military program, Integrated Cognitive-Neuroscience Architectures for
Sensemaking (ICArUS), "geospatial sense making" merges cranial and
geological space. The ICArUS program is focused on "models of geospace
whose functional architecture conforms closely with that of the human
brain."[164] Neural signals of target detection have been observed at rates of
twenty images per second. It is not feasible to match this rapid rate of visual
input in the brain with the slow motor response of pressing a button, which
takes three hundred microseconds, depending on the dexterity of the button pusher. Linking response to a target as closely as possible to its visual

representation in the brain would compress the kill chain substantially. What Hayles calls the "potent force of human and technical systems"[165] gains potency once combined because machinic and human assemblages speed up and expand military violence.

Memory, especially traumatic memory, is where a good deal of war neuroscience is happening.[166] But the discovery of cognitive nonconsciousness also introduces the way in which thought itself can be activated as a weapon. Cognitive firing can become a more fully kinetic kind of enterprise than it already is. Thus the epoch of cognitive avionics is upon us. War pilots are able to use "halo sport brain stimulators to accelerate strength and skill gains by up to half."[167] According to DARPA's Next-Generation Nonsurgical Neurotechnology (N3), human–machine interactions are being bent toward "defense relevant tasks that can read from and write to 16 independent channels within a 16mm to the third power volume of neural tissue within 50ms."[168] The move from enhancing a soldier's performance in the form of psychoactive pills to the fighter jet itself as a "cognitive neural prosthetic" bypasses the psyche altogether.[169] In this instance, the plane, missile, rifle, or small arms device is an extension of the cognitive state. The wireless linking between a pilot's brain and the enhanced video-vision of drones is now an operational option. Innovations in noninvasive brain-machine interfaces have turned into telepathic helmets.[170] As one headline reads, "Mind-Controlled Drones Race to the Future."[171]

Another example of a war-directed brain-machine interface is DARPA's cyber-beetle. Here a living insect is implanted with miniature hardware enabling human arachno-pilots to control the muscle systems necessary to fly the bug by wire.[172] To capture the radical (DARPA's term) degree to which war is changing fundamental premises about the human being per se, it is not enough to point out the weaponization of identity and culture. Posthuman war is now a state of mind. To cite Connolly's final assessment: "It is possible to imagine a species, something like us, with quick capacities of explicit memory retrieval, and parallel processing reserved for virtual memory prior to consciousness, transcending thereby the crude limits of the human."[173] Transcending the limits of the human is not only possible, it is certain: war neuroscience is moving the evolution of the species in Connolly's heady direction. It is doing so beginning with the HTS program and ending in the *terrain*-ing of cognition itself—the human being is reconceived as both matter and system. But the story does not stop at recombining subjects and objects in this way. In future work I will provide further analysis on the arming of matter at the atmospheric level, with a focus on environmental modification and posthuman war.

Notes

INTRODUCTION

1. Malcolm W. Nance, *Terrorist Recognition Handbook*, 2nd ed. (Boca Raton, Fla.: Taylor and Francis, 2008), 10.

2. Donald J. Trump, *National Security Strategy of the United States of America*, December 2017, https://www.acq.osd.mil/ncbdp/docs/NSS-Final-12-18-2017-0905.pdf.

3. The human cost of post-9/11 wars in terms of direct war deaths is more than eight hundred thousand, over three hundred thousand of them civilian. For a more specific accounting, see the Costs of War Project, https://watson.brown.edu/costsofwar/files/cow/imce/papers/2019/Direct%20War%20Deaths%20COW%20Estimate%20November%2013%202019%20FINAL.pdf. For the number of U.S. dead and wounded since the Revolutionary War, see U.S. Department of Veterans Affairs, "America's Wars," December 2019, https://www.va.gov/opa/publications/factsheets/fs_americas_wars.pdf.

4. Sydney J. Freedberg, "Bullets, Beans, and Data: The New Army Materiel," *Breaking Defense*, December 15, 2020, https://breakingdefense.com/2020/12/bullets-beans-data-the-new-army-materiel-command-exclusive/.

5. The term "warrior intellectuals" is associated with General Petraeus in Roberto J. González, "Toward Mercenary Anthropology," *Anthropology Today* 23, no. 3 (June 2007): 17.

6. Department of the Army, *Army Special Operations Manual on Unconventional Warfare* (FM 3–05.130) (Washington, D.C.: Department of the Army, 2008), 11.

7. Foucault introduces the concept of biopower in *Society Must Be Defended* (New York: Picador, 2003). On biopower elsewhere in Foucault, see *The History of Sexuality: An Introduction, vol. 1* (New York: Vintage, 1980).

8. Michel Foucault, *The Foucault Effect: Studies in Governmentality*, ed.

Graham Burchell, Colin Gordon, and Peter Miller (London: University of Chicago Press, 1991), 36.

9. On the term World War X, see Robert M. Cassidy, *Counterinsurgency and the Global War on Terror: Military Culture and Irregular War* (Stanford, Calif.: Stanford Security Studies, 2008), 1–20.

10. Robert Esposito, *Immunitas: The Protection and Negation of Life* (Cambridge, U.K.: Polity Press, 2011).

11. David Kilcullen, *The Accidental Guerrilla: Fighting Small Wars in the Midst of a Big One* (Oxford: Oxford University Press, 2009).

12. Judith Butler, *Precarious Lives: The Powers of Mourning and Violence* (London: Verso, 2004), xv.

13. Georgio Agamben, *State of Exception* (Chicago: University of Chicago Press, 2005).

14. Brian Massumi, *Ontopower: War, Powers, and the State of Perception* (Durham, N.C.: Duke University Press, 2015), 5.

15. Bernard Stiegler, *Technics and Time, 1: The Fault of Epimetheus*, trans. Richard Beardsworth and George Collins (Palo Alto, Calif.: Stanford University Press, 1998).

16. Department of Defense, *The Department of Defense Net-Centric Data Strategy* (Washington, D.C.: Department of Defense, 2003), ii.

17. On war as a duel on a larger scale, see Claus von Clausewitz, *On War* (London: Oxford University Press, [1832] 2007), 12–15.

18. DOD, *Net-Centric Data Strategy,* 9.

19. DOD, 5.

20. Norbert Wiener, *Cybernetics, or Control and Communication in the Animal and Machine,* 2nd ed. (Cambridge, Mass.: MIT Press, [1948] 1965).

21. For a critique of numerical measurement as an escape from so-called real complexity, see Dean Pieridies, Michael J. Zypher, and Jon Roffe, "Measurement and Statistics in 'Organization Science': Philosophical, Sociological, and Historical Perspectives," in *The Routledge Companion to Philosophy in Organizational Studies,* ed. Raza Mir, Hugh Willmott, and Michelle Greenwood, 474–82 (New York: Routledge, 2015). On counting as a "control technology," see Michael Power, "Counting, Control, and Calculation: Reflections on Measuring and Management," in *Human Relations* 57, no. 6 (2004): 765–83. For more on the critique of digital technology, see Nick Dyer-Witheford, *Cyber-Proletariat: Global Labour in the Digital Vortex* (London: Pluto Press, 2015); and Ursula Huws, *The Cyberteriat Comes of Age: Labor in the Global Digital Economy* (New York: Monthly Review Press, 2014).

22. This reference to netcentrism as being in opposition to democracy is

from Ned Rossiter, "Organized Networks and Nonrepresentative Democracy," in *Reformatting Politics: Information, Technology, and Global Civil Society*, ed. Jodi Dean, Jon W. Anderson, and Geert Lovnik, 49–64 (New York: Routledge, 2006).

23. Zypher and Roffe, "Measurement and Statistics," 478.

24. Bruno Latour, *We Have Never Been Modern* (Cambridge, Mass.: Harvard University Press, 1993), 35, 43.

25. Manuel DeLanda, *War in the Age of Intelligent Machines* (New York: Zone Books, 1991).

26. Defense Advanced Research Projects Agency (DARPA), "ARPANET and the Origins of the Internet," https://www.darpa.mil/about-us/timeline/arpanet, unpaginated. On the transition from ARPANET to the internet, see Janet Abbate, *Inventing the Internet* (Cambridge, Mass.: MIT Press, 1999).

27. Shane Harris, *@War: The Rise of the Military-Internet Complex* (New York: Houghton Mifflin Harcourt, 2014).

28. United Nations, "Article 42," *Universal Declaration of Human Rights*, http://www.un.org/en/udhrbook/pdf/udhr_booklet_en_web.pdf, 50.

29. Paul Virilio, *The Information Bomb* (London: Verso, 2006).

30. Félix Guattari, *Chaosmosis: An Ethico-Aesthetic Paradigm*, trans. Paul Bains and Julian Pefanis (Bloomington: Indiana University Press, 1995), 35.

31. For more on the technologization of human bodies and minds without the good or evil mandate, see Elaine L. Graham, *Representations of the Post/Human: Monsters, Aliens, and Others in Popular Culture* (New Brunswick, N.J.: Rutgers University Press, 2002).

32. See David Deutsch, *The Fabric of the Reality: The Science of Parallel Universes and Its Implications* (London: Penguin, 1998); and Max Tegmark, *Our Mathematical Universe: My Quest for the Ultimate Nature of Reality* (New York: Knopf, 2014).

33. Donna Haraway, "A Cyborg Manifesto: Science, Technology, and Socialist Feminism in the Late Twentieth Century," in *Literary Criticism: Literary and Cultural Studies*, 4th ed., ed. Robert Con Davis and Ronald Schleifer, 695–727 (New York: Longman, 1998).

34. Ronald Cole-Turner, *The New Genesis: Theology and the Genetic Revolution* (Louisville: Westminster/John Knox Press, 1993), 9.

35. On prosthetic theory and human–machine hybrids, see David Wills, *Prosthesis* (Stanford, Calif.: Stanford University Press, 1993); and Gabriel Brahm Jr. and Mark Driscoll, eds., *Prosthetic Territories: Politics and Hypertechnologies* (Boulder: Westview Press, 1995). On prostheses as central to posthumanism, see Marquaard Smith and Joanne Morra, eds., *The Prosthetic Impulse: From a Posthuman Present to a Biocultural Future* (Cambridge, Mass.: MIT Press, 2006).

36. Pramod K. Nayar, *Posthumanism* (Cambridge, U.K.: Polity Press, 2014), 5.

37. Nayar, 5.

38. Timothy Morton, *The Ecological Thought* (Cambridge, Mass.: Harvard University Press, 2010). Hereafter referred to by page number in text as *ET*.

39. Morton, 3.

40. Donna Haraway, *Staying with the Trouble: Making Kin in the Chthulucene* (Durham, N.C.: Duke University Press, 2016), 33.

41. Cary Wolfe, *What Is Posthumanism?* (Minneapolis: University of Minnesota Press, 2010), xviii.

42. Haraway, *Staying with the Trouble*, 11.

43. Graham Harman, "The Well-Wrought Broken Hammer," *New Literary History* 43, no. 2 (Spring 2012): 187. For more on Harman's version of realism, see Manuel DeLanda and Graham Harman, *The Rise of Realism* (Cambridge, U.K.: Polity Press, 2017). For a study of antirealism as the predominate strain of continental Western philosophy after Kant, see Lee Braver, *A Thing of This World: A History of Continental Anti-realism* (Evanston, Ill.: Northwestern University Press, 2007). On the relation between realism and more traditional forms of historical materialism, see Geoff Pfeifer, *The New Materialism: Althusser, Badiou, and Zizek* (London: Routledge, 2015).

44. Jane Bennett, "Systems and Things: A Response to Graham Harman and Timothy Morton, *New Literary History* 43 (2012): 225–33. See also Graham Harman, "I Am Also of the Opinion That Materialism Must Be Destroyed," *Society and Space* 28 (2010): 772–90.

45. Jane Bennett, "Systems and Things: On Vital Materialism and Object-Oriented Philosophy," in *The Non-Human Turn*, ed. Richard Grusin (Minneapolis: University of Minnesota Press, 2015), 231.

46. Bennett, 226.

47. Bennett, 228.

48. For a full account of the origins of system thinking in the Enlightenment, see Clifford Siskin, *System: The Shaping of Modern Knowledge* (Cambridge, Mass.: MIT Press, 2016). Siskin is against using the word "system" as a catch-all phrase, or a way to play what he calls "the blame game" (165–68). I've noted already that this is a game I would also like to avoid. Posthuman war seeks to make effective use of system theory, but not all systems are inherently connected to war.

49. Office of the Secretary of Defense, *Systems Engineering Guide for Systems of Systems* (Washington, D.C.: Department of Defense, 2008).

50. Siskin, *System*, 231, 233.

51. Jeffrey Engstrom, *Systems Confrontation and System Destruction Warfare: How the Chinese People's Liberation Army Seeks to Wage Modern Warfare* (Santa Monica, Calif.: RAND, 2018).

52. Timothy Morton, *Hyperobjects: Philosophy and Ecology after the End of the World* (Minneapolis: University of Minnesota Press, 2013), 18.

53. Bruno Latour, *Facing Gaia: Eight Lectures on the New Climactic Regime* (Cambridge, U.K.: Polity Press, 2017), 223.

54. Bruno Latour, *We Have Never Been Modern* (Cambridge, Mass.: Harvard University Press, 1993). On becoming "terrestrialized," see Latour, *Coming Down to Earth: Politics in the New Climate Regime* (Cambridge, U.K.: Polity Press, 2018).

55. Latour, 45.

56. Latour, *Facing Gaia*, 223.

57. Latour, 223, emphasis in original.

58. On the democracy of things, see Bruno Latour, *Politics of Nature: How to Bring the Sciences into Democracy* (Cambridge, Mass.: Harvard University Press, 2005).

59. Here again I refer the reader to earlier work on so-called postwhiteness in my previous book, *After Whiteness*.

60. For an example of the "flea" metaphor in counterinsurgency theory, see Robert Taber's classic COIN text, *The War of the Flea: A Study of Guerrilla Warfare Theory and Practice* (New York: Citadel Press, 1965).

61. For a general accounting of the HTS program, with all the facts and figures on funding, personnel, and timelines, see Catherine Lutz, "The Military Normal," in *The Counter-Counterinsurgency Field Manual: Or, Notes on the Demilitarization of American Society,* ed. Network of Concerned Anthropologists, 23–38 (Chicago: Prickly Paradigm Press, 2009).

62. Gregory Bateson, *Steps to an Ecology of Mind* (Chicago: University of Chicago Press, [1972] 2000).

63. The term "material semiotic" is borrowed from Carolin Gerlitz and Celia Lury's "Social Media and Self-Evaluating Assemblages: On Numbers, Orderings, and Values," *Distinktion: Journal of Social Theory* 15, no. 2 (2014): 1.

64. R. Douglas Fields, "White Matter," *Scientific American* (March 2008), 55.

65. Cornelia Bargman, et al., *BRAIN 2025: A Scientific Vision*, 8, 9, 16, https://www.ninds.nih.gov/sites/default/files/brain2025_508c.pdf .

66. Bargman et al., 20, 46.

67. Bargman et al., 14.

68. Levi R. Bryant, *Onto-Cartography: An Ontology of Machines and Media* (Edinburgh: Edinburgh University Press, 2014).

69. National Research Council, *Opportunities in Biotechnology for Future Army Applications* (Washington, D.C.: National Academy Press, 2001), 3.

70. Sven Lindqvist, *A History of Bombing* (New York: New Press, 2003).

71. Wiener, *Cybernetics*, 2.

72. Wiener, 29.

73. Gilles Deleuze, "Control and Becoming," in *Negotiations* (New York: Columbia University Press, 1995), 171.

1. WAR DEMOGRAPHY

1. Robert Gates, "American Association of Universities," Monday, April 14, 2008, http://www.defenselink.mil/speeches/speech.aspx?speechid=1228.

2. Paula Holmes-Eber, *Culture in Conflict: Irregular Warfare, Culture Policy, and the Marines Corps* (Stanford, Calif.: Stanford Security Studies, Stanford University Press, 2014), 142.

3. Michael D. Matthews, *Head Strong: How Psychology Is Revolutionizing War* (Oxford: Oxford University Press, 2014), 6, 93.

4. "Message from the Secretary Chief of Staff and Sergeant Major of the Army," in *United States Army Diversity Road Map,* December 2015, http://www.armydiversity.army.mil/document/diversity_roadmap.pdf.

5. Matthews, *Head Strong,* 112.

6. See "Army Demographics," https://m.goarmy.com/content/dam/go army/downloaded_assets/pdfs/advocates-demographics.pdf.

7. Gates, "American Association of Universities," 3. The Minerva Initiative seeks to bring together "a consortia of universities" to provide open-source documentary archives of cultures that garner military interest.

8. Gates, 1. The Minerva project was budgeted for approximately $100 million over five years in 2008. See Sharon Weinberger, "Military Research: The Pentagon's Culture Wars," *Nature* 455 (October 2008): 583–85.

9. The most famous example is the CORDS project in Vietnam, which has been linked to the assassination of more than twenty-six thousand Vietcong. See David Vine, "Enabling the Kill Chain," *Chronicle of Higher Education,* November 30, 2007, B9.

10. For more on the RMA, see Thomas K. Adams, *The Army after Next: The First Postindustrial Army* (Westport, Conn.: Praeger, 2006).

11. Gates, "American Association of Universities," 1.

12. Gates, 3.

13. Robert Gates, "US Global Leadership Campaign," July 15, 2008, https://smallwarsjournal.com/blog/secretary-gates-at-the-us-global-leadership -campaign.

14. Gates, 5.

15. Gates, 6. On the essential connections between war and the university in the United States, see Robin W. Winks, *Cloak and Gown: Scholars in the Secret War, 1939–1961* (New York: William Morrow, 1987); beyond that period, see David Montgomery, ed., *The Cold War and the University* (New York: New Press, 1997).

16. Michel Foucault, *Society Must Be Defended: Lectures at the College de France,* trans. David Macy, 1975–76 (New York: Picador, 1977), 16.

17. George W. Bush, *US National Security Strategy* (Washington, D.C.: The White House, 2002).

18. Jürgen Habermas, *The Structural Transformation of the Public Sphere* (Cambridge, Mass.: MIT), 29.

19. See Paul Virilio, *Pure War* (New York: Semiotexte, 1997).

20. See Joe Halderman, *The Forever War* (New York: Saint Martins, 1974).

21. The American Civil Liberties Union, *War Comes Home: The Excessive Militarization of American Policing* (New York: ACLU, 2014), 23.

22. Alain Badiou, *Briefings on Existence: A Short Treatise on Transitory Ontology,* trans. Norman Madarasz (Albany: State University of New York Press, 2006), 36–37. For more on Badiou and war, see Nick Mansfield, *Theorizing War: From Hobbes to Badiou* (London: Palgrave Macmillan, 2008).

23. Badiou, 37.

24. Alain Badiou, *Number and Numbers* (Cambridge, U.K.: Polity Press, 2008), 1.

25. Carl Schmitt, *The Crisis of Parliamentary Democracy,* trans. Ellen Kennedy (Cambridge, Mass.: MIT Press, 1986).

26. For more on the U.S. census, see Mike Hill, *After Whiteness: Unmaking an American Majority* (New York: New York University Press, 2004).

27. Office of Management and Budget (OMB), "Statistical Policy Directive No. 15," 1977, https://clintonwhitehouse2.archives.gov/omb/fedreg/notice_15.html.

28. In *The Monthly Labor Review,* a standard source of debate on federal statistical policies, Ruth B. Mckay and Manuel de la Puente report an increased public sense of "confusion" and "suspicion" regarding the U.S. census. See their article, "Cognitive Testing of Racial and Ethnic Questions for the CPS Supplement," *MLR* (September 1996): 8–12. On the general fragility of current race categories, see Lawrence Wright's "One Drop of Blood," *New Yorker* (July 25, 1994): 46–55; and Dennis Barron, "How to Be a Person, Not a Number," *Chronicle of Higher Education,* April 3, 1998, B-8. James F. Davis's *Who Is Black: One Nation's Definition* (University Park: Pennsylvania State Press, 1991) remains one of the most trenchant critiques of the ironic twists and turns of the enforcement of the one drop rule of black hypodescent. See also Michael Omi, "Racial Identity and the State: The Dilemmas of Classification," *Law and Inequality* 15, no. 7 (1997): 14–28.

29. Calvert Dedrick, head of the federal Statistical Research Division, was sent to California in 1942 to supervise the statistical work necessary for the internment of innocent Japanese. See Margo Anderson, *The American Census: A Social History* (New Haven, Conn.: Yale University Press, 1990), 194.

30. Reginald Daniel, *More Than Black: Multiracial Identity and the New Racial Order* (Philadelphia: Temple University Press, 2002), 3.

31. Daniel, 13.

32. In the 1850 census the term "mulatto" was used and in 1890 populations could be identified as "quadroon" and "octoroon." The new decision is not substantially different from the interagency recommendations issued in the 1977 mandates of OMB 15. For the census 2000 standards, see *Federal Register 62*, no. 131 (July 9, 1997).

33. See William O'Hare, "Managing Multiple-Race Data," in *American Demographics* (April 1998): 44.

34. In a letter presented at the congressional hearings in 1993, the nation's top civil rights leaders expressed "extreme concern that the new [multiracial] category will inadvertently cause confusion and inconsistent reporting." See *Hearings before the Subcommittee on Census, Statistics, and Postal Personnel,* serial no. 103-7, April 4, 1993 (Washington, D.C.: U.S. Government Printing Office), 224. An editorial by Charles Byrd, "The Political Realignment: A Jihad Against 'Race' Consciousness," *Interracial Voice,* September 5, 2000, blames the NAACP directly for maintaining "the one drop rule" and discouraging multiracial census reclassification. *Interracial Voice* is found online at https://www.interracialvoice.com.

35. Morefield Story, cited in Mary L. Dudziak, *Cold War Civil Rights: Race and the Image of American Democracy* (Princeton, N.J.: Princeton University Press, 2011), 7.

36. Jami L. Bryan, "Fighting for Respect: African-American Soldiers in WWI," Historical Army Foundation, https://armyhistory.org/fighting-for-respect-african-american-soldiers-in-wwi/.

37. Cited in Dudziak, 86.

38. Malcolm W. Nance, *Terrorist Recognition Handbook,* 2nd ed. (Boca Raton, Fla.: Taylor and Francis, 2008), 32.

39. Jürgen Habermas, *The Inclusion of the Other* (Cambridge, Mass.: MIT Press, 1998), 113.

40. Jürgen Habermas, "Remarks on Legitimation through Human Rights," *Modern Schoolman 75,* no. 2 (January 1998), 87–100.

41. Quentin Meillassoux, *After Finitude: An Essay on the Necessity of Contingency* (London: Bloomsbury, 2006), 5–6.

42. For more on Cantor's relation to Meillassoux, see Paul Livingston, "Cantor, Georg," in *The Meillassoux Dictionary,* ed. Peter Graton and Paul J. Ennis, 37–40 (Edinburgh: Edinburgh University Press, 2014).

43. Alain Badiou, *Being and Event* (London: Bloomsbury, 2007), 280.

44. Badiou, 391.

45. Alain Badiou, *Theory of the Subject,* trans. Bruno Bosteels (London: Bloomsbury), 84.

46. Alain Badiou, *Number and Numbers,* trans. Robin MacKay (Cambridge, U.K.: Polity Press, 2008), 211.

47. Headquarters, *Department of the Army, Special Forces Guide,* TC 31–73 (Washington, D.C.: Department of the Army, 2008), 28.

48. Quentin Meillassoux, "Potentiality and Virtuality," *Collapse* 2 (2007): 55–81, 230.

49. Meillassoux, "Potentiality and Virtuality," 232.

50. Badiou, *Being and Event,* 279.

51. Badiou, 27.

52. Gregoire Chamayou, *A Theory of the Drone* (New York: New Press, 2015), 146–48.

53. On what they call the "withering of civil society," see Michael Hardt and Antonio Negri, *The Labor of Dionysus: A Critique of State-Form* (Minneapolis: University of Minnesota Press, 1994).

54. Jürgen Habermas, "Kant's Idea of Perpetual Peace, with the Benefit of Two Hundred Years' Hindsight," in *Perpetual Peace: Essays on Kant's Cosmopolitan Ideal,* ed. James Bohman and Matthias Lutz-Bachmann, 111–45 (Cambridge, Mass.: MIT Press, 1997.

55. Habermas, 115.

56. Immanuel Kant, *The Metaphysics of Morals,* trans. J. W. Semple (Edinburgh: T&T Clark, 1886), 496. Page numbers refer to the Online Library of Liberty, https://oll.libertyfund.org/title/calderwood-the-metaphysics-of-ethics.

57. On Kant's opposition to the right of resistance, see Reider Malik, *Kant's Politics in Context* (Oxford: Oxford University Press, 2014).

58. On transcendental dualisms in Kant, see Allen W. Wood, "Introduction," in *The Cambridge History of Philosophy in the Nineteenth Century (1790–1870),* ed. Allen W. Wood, 1–9 (Cambridge: Cambridge University Press, 2018).

59. On property as an aquired right in Kant, see Nelson Potter, "Applying the Categorical Imperative in Kant's *Rechtslehre,*" *Criminal Law and Legal Philosophy* (2003): 37–51.

60. See Elias Canetti, *Auto-da-Fé* (New York: Farrar, Straus and Giroux, 1984).

61. For more on Kant's detachment from the phenomenal world, see Katrin Flikschuh, "On Kant's *Rechtslehre,*" *European Journal of Philosophy* 5, no. 1 (1997): 50–73; for Kant's rationalism as positioned too far from nature, see Peter Critchley, "Kant's Natural Teleology and Moral Praxis," *Humanities Commons* (2012), http://dx.doi.org/10.17613/M67S7HR8T.

62. Badiou, *Number and Numbers*, 1–2.

63. Badiou, xvii.

64. Immanuel Kant, *Political Writings* (Cambridge: Cambridge University Press, 1991), 114 ff.

65. The literature on Enlightenment riots is vast, but E. P. Thompson's work is the locus classicus.

66. On the multitude and the Enlightenment, see Mike Hill and Warren Montag, *The Other Adam Smith* (Stanford, Calif.: Stanford University Press, 2015).

67. For a sustained response to this contradiction over numbers as "masses," see the essays in *Masses, Classes, and the Public Sphere*, ed. Mike Hill and Warren Montag (London: Verso, 2001), especially the introduction.

68. Habermas, *Inclusion of the Other*, 177.

69. Habermas, 176.

70. Habermas, *Structural Transformation of the Public Sphere*, 43–44.

71. For a recent turn in the quantitative direction, see Franco Moretti, *Distance Reading* (London: Verso, 2013), 45–46.

72. Immanuel Kant, "Perpetual Peace: A Philosophical Sketch," in *Kant: Political Writings*, ed. Hans Reiss (Cambridge: Cambridge University Press), 93–129.

73. Thomas Hobbes, *On the Citizen* (Cambridge: Cambridge University Press, [1642] 2003), xl, xlii. On the terms "constitutive" and "constituent" power, see Antonio Negri, *The Dionysus of Labor* (Minneapolis: University of Minnesota Press, 1994), 283–85.

74. Alain Badiou, *Mathematics of the Transcendental* (London: Bloomsbury, 2014), 29.

75. These estimates are taken, respectively, from Will Kymicka, *Multicultural Citizenship: A Liberal Theory of Minority Rights* (Oxford: Clarendon Press, 1995), 1; and James Tully, *Strange Multiplicity: Constitutionalism in an Age of Diversity* (Cambridge: Cambridge University Press, 1995), 3.

76. Badiou, *Being and Event*, 283.

77. See Carl Schmitt, *Theory of the Partisan* (New York: Telos Publishing, 2007).

78. Meillassoux, *After Finitude*, 57.

79. Badiou, *Being and Event*, 64.

80. Kant, *Perpetual Peace*, 114.

81. Kant, 114, emphasis added.

82. Habermas, *Inclusion of the Other*, 176.

83. Kant, *Perpetual Peace*, 114.

84. On Foucault's turn to the discourse of rights, see Ben Golder, *Foucault and the Politics of Rights* (Stanford, Calif.: Stanford University Press, 2015).

85. Leveler, for example, was a derogatory term dating back to the Midland Revolts of 1607 against the Enclosure Acts, referring to those who leveled the hedges meant as living fences to keep subsistence farmers away from formerly common land. On Puritanism as protocommunism and the English Civil War, see James Holstun, *Ehud's Dagger: Class Struggle in the English Revolution* (London: Verso, 2000).

86. The first English departments were established in places like the Scottish Highlands and British India, for the reason of establishing English-ness in both the national and disciplinary senses of that term. On English literature in Scotland, see Robert Crawford, *The Scottish Invention of English Literature* (Cambridge: Cambridge University Press, 1998). On literary education in India, see Gauri Viswanathan, *Masks of Conquest: Literary Study and British Rule in India* (New York: Columbia University Press, 1989).

87. Wolfgang Ernst, *Digital Memory and the Archive* (Minneapolis: University of Minnesota Press, 2013), 55.

2. WAR ANTHROPOLOGY

1. For a general accounting of the HTS program, with all the facts and figures on funding, personnel, and time lines, see Catherine Lutz, "The Military Normal," in *The Counter-Counterinsurgency Field Manual: Or, Notes on the Demilitarization of American Society,* ed. Network of Concerned Anthropologists (Chicago: Prickly Paradigm Press, 2009), 23–38.

2. David Price, *AAA Commission on the Engagement of Anthropology with the US Security and Intelligence Communities,* November 4, 2007, 19.

3. For an overview of ethical considerations in the military uses of anthropology ranging from Vietnam to HTS, see George R. Lucas Jr., *Anthropologists in Arms: The Ethics of Military Anthropology* (New York: Rowman and Littlefield, 2009). A complete version of the AAA's "Code of Ethics" appears in the appendix.

4. Robert F. Kennedy, cited in Michael McClintock, *Instruments of Statecraft: U. S. Guerrilla Warfare, Counter-insurgency, and Counter-terrorism, 1940–1990* (New York: Pantheon Books, 23).

5. The term "warrior intellectuals" is associated with General Patraeus in Roberto J. González, "Toward Mercenary Anthropology," *Anthropology Today* 23, no. 3 (June 2007): 17.

6. On insurgency in the American Revolutionary War, see Walter E. Kretchik, *U.S. Army Doctrine: From the American Revolution to the War on Terror* (Lawrence: University Press of Kansas, 2011); on the difference between U.S. and British tactics during this peroid, see Jeremy Black, *War and the Cultural Turn*

(Cambridge, U.K.: Polity Press, 2011); on the U.S. Civil War, see Mark Moyar, *A Question of Command: Counterinsurgency from the Civil War to Iraq* (New Haven, Conn.: Yale University Press, 2009).

7. *U. S. Counterinsurgency Manual FM31–15* (Washington, D.C.: Department of the Army, 1961), https://archive.org/details/FM31-151961.

8. Seymour J. Deitchman, *The Best Laid Schemes: A Tale of Social Research and Bureaucracy* (Quantico, Va.: Marine Corps University Press, [1976] 2014), 182.

9. Fred Eggan, "Annual Meeting, December 30, 1942, 9:00 AM," *American Anthropologist* 44, no. 1 (1942): 634–37.

10. Franz Boas, "Scientists as Spies," *Nation*, December 20, 1919: 797.

11. Boas, 797.

12. Deitchman, *Best Laid Schemes*, 183.

13. Margaret Mead, "Preface—1965," in *And Keep Your Powder Dry: An Anthropologist Looks at America*, xxxi–xxxiii (New York: William Morrow, [1942] 1965), xxxi.

14. Margaret Mead, "The Comparative Study of Culture and the Purposive Cultivation of Democratic Values, 1941–1949," in *Perspectives on a Troubled Decade: Science, Philosophy, Religion, 1939–1949*, ed. Lyman Bryson et al., 81–93 (New York: Conference on Science, 1950), 88–89.

15. Richard Stubbs, *Hearts and Minds in Guerrilla Warfare: The Malayan Emergency, 1948–1960* (Oxford: Oxford University Press, 1989), 1.

16. Stubbs, *Hearts and Minds*, 1.

17. Bronislaw Malinoski, "Practical Anthropology," in *Africa* 2, no. 1 (1929): 22–38. For a critique of this and of war anthropology in general, see again David Price, *Weaponizing Anthropology: Social Science in Service of the National Security State* (Chico, Calif.: AK Press, 2011).

18. Thomas Jefferson to Captain Lewis Meriwether, June 20, 1803, https://encyclopediavirginia.org/entries/thomas-jeffersons-instructions-to-meriwether -lewis-june-20-1803/. On the origins of COIN doctrine in the 1800s, see Thomas Rid, "The Nineteenth-Century Origins of Counterinsurgency Doctrine," *Journal of Strategic Studies* 33, no. 5 (October 2011): 727–58.

19. For more on the military aspects of Lewis and Clark's expedition, see Price, *Weaponizing Anthropology*.

20. Samuel Huntington, *The Soldier and the State: The Theory and Politics of Civil-Military Relations* (Cambridge, Mass.: Harvard University Press, 1957), 229. On insurgency practices in the Civil War era, see Robert R. Mackey, *The Uncivil War: Irregular Warfare in the Upper South, 1861–1865* (Norman: University of Oklahoma Press, 2004).

21. John Wesley Powell, in his testimony to Congress, cited by Curtis M.

Hinsely Jr., "Anthropology as Science and Politics: The Dilemmas of the Bureau of American Ethnology, 1879 to 1904," in *The Uses of Anthropology,* ed. William Goldschmidt, 15–32 (Washington, D.C.: American Anthropological Association, 1979), 18.

22. Powell, cited in Hinsely Jr., "Anthropology as Science and Politics," 19.

23. For an excellent accounting of Mead's involvement in the cultural turn in war before and leading up to World War II, see David Price, *Anthropological Intelligence: The Deployment and Neglect of American Anthropology in the Second World War* (Durham, N.C.: Duke University Press, 2008), 19ff.

24. Margaret Mead, "The Uses of Anthropology in World War II and After," in *The Uses of Anthropology,* ed. William Goldschmidt, 145–57 (Washington, D.C.: American Anthropology Association, 1979), 146.

25. Mead, *And Keep Your Powder Dry,* 21, emphasis in original.

26. Margaret Mead and Rhoda Metraux, eds., *The Study of Culture at a Distance,* vol. 1 (New York: Berghan Books, [1949] 2000), 3.

27. Mead, *And Keep Your Powder Dry,* 16.

28. Special Services Division, United States Army, *A Pocket Guide to West Africa* (Washington, D.C.: War and Navy Departments, 1943), 11.

29. On the condom in war, see Stephen L. Harp, *A World History of Rubber: Empire, Industry, and the Everyday* (Chichester, U.K.: Wiley-Blackwell, 2015).

30. U.S. Marine Corps, *Small Wars Manual,* FMFRP 12–15 (1940), 19. On the ideology of "prophylaxis," see Mike Hill, "Can Whiteness Speak? Institutional Anomies, Ontological Disasters, and Three Hollywood Films," in *White Trash: Race and Class in America,* ed. Annallee Newitz and Matt Wray, 155–76 (New York: Routledge, 1997).

31. Stubbs, *Hearts and Minds,* 181.

32. The MRLA did not represent the Malayan "races" in any real sense, since its membership was over 95 percent Chinese. Their significance as an opposing force to the British and Malayan governments lie in earlier communist resistance to the Japanese. See Walter C. Ladwig III, "Managing Counterinsurgency: Lessons from Malaya," *Military Review,* May–June 2007, 56–68.

33. S. F. Nadel, *The Foundations of Social Anthropology* (London: Cohen and West, 1953), 2.

34. Bronislaw Malinowski, "The Rationalization of Anthropology and Administration," *Africa* 3 (1930), 4.

35. Malinowski, 145, 160.

36. On losing the irregular war in Vietnam, see John Nagl, *Counterinsurgency Lessons from Malaya and Vietnam: Learning to Eat Soup with a Knife* (Westport, Conn.: Praeger, 2002); Deborah D. Avant, *Political Institutions and Military*

Change: Lessons from Peripheral Wars (Ithaca: Cornell University Press, 1994); and R. D. Downie, *Learning from Conflict: The US Military in Vietnam, El Salvador, and the Drug War* (Westport, Conn.: Praeger, 1998).

37. H. G. Summers, *On Strategy: A Critical Analysis of the Vietnam War* (Navato, Calif.: Presidio Press, 1982).

38. These numbers come from Robert Egnell, *Complex Peace Operations and Civil-Military Relations* (New York: Routledge, 2009), 43.

39. On the failure of census legibility in the Strategic Hamlet program, see James Scott, *Seeing Like a State* (New Haven, Conn.: Yale University Press, 1998).

40. Andrew Mumford, *Puncturing the Counterinsurgency Myth: Britain and Irregular Warfare in Past, Present, and Future* (Carlisle Barracks, Penn.: US Army War College, Strategic Studies Institute, 2011), 1, 21.

41. Walter Ong, "Ramus: Rhetoric and the Pre-Newtonian Mind," in *An Ong Reader: Challenges for Further Inquiry*, ed. Thomas J. Farrel and Paul A. Soukup, 209–38 (New Jersey: Hampton Press, [1954] 2002).

42. Ong, 219.

43. Early copyright law in the eighteenth century focused on what the adjudicators called "the de-corporealization of the idea." For an analysis of this term and this history, see Mike Hill and Warren Montag, *The Other Adam Smith* (Stanford, Calif.: Stanford University Press, 2014), chapter 3.

44. Ong, "Written Transmission of Literature," in *An Ong Reader*, 331–44.

45. Ong, "Digitization, Ancient and Modern," in *An Ong Reader*, 527.

46. Ong, 528.

47. Ong, "Digitization, Ancient and Modern," 528.

48. Ong, 541.

49. Ong, "Information and/or Communication," in *An Ong Reader*, 508.

50. Ong, 515.

51. Ong, "Ramus," 214, 234–35.

52. Ong, 226.

53. U.S. Marine Corps, *Small War Manual*, 24.

54. U.S. Marine Corps, 24.

55. T. E. Lawrence, "The Evolution of a Revolt," *Army Quarterly and Defense Journal* (October 1920), https://permanent.fdlp.gov/lps68452/lawrence.pdf.

56. Patrick Porter, "Good Anthropology, Bad History: The Cultural Turn in Studying War," *Parameters* (Summer 2007): 46. For similar cautions against war's cultural return, see Robert H. Scales Jr., "Culture-Centric Warfare," *Proceedings* 130 (October 2004), 32–36; and George Packer, "Knowing the Enemy: Can Social Scientists Redefine the 'War on Terror?,'" *New Yorker*, December 18, 2006, 60–69.

57. Eliza Jane Darling, "The Anthropology Wars," in *Monthly Review* 64, no. 7 (December 2012): 3.

58. On World War II anthropology, see Rudolf V. A. Janssens, "Toilet Training, Shame, and the Influence of Alien Cultures: Cultural Anthropologists and American Policy Making in Postwar Japan, 1944–45," in *Anthropology and Colonialism in Asia and Oceania*, ed. Jan van Bremen, 285–307 (Surrey, U.K.: Curzon Press, 1999).

59. Cited in Price, *Anthropological Intelligence*, 37.

60. For a particularly good essay on biological weapons imagined for use against the Japanese and their relation to cultural anthropology, see David Price, "How US Anthropologists Planned 'Race Specific' Weapons against the Japanese," *Counterpunch*, November 25, 2005, http://www.counterpunch.org/2005/11/25/how-us-anthropoligists-planned-quot-race-specific-quot-weapons-against-the-japanese/.

61. Price, 5.

62. Price, 5.

63. Cited in Price, *Anthropological Intelligence*, 225.

64. David H. Price, "Anthropology and Total Warfare: The Office of Strategic Services' 'Preliminary Report on Japanese Anthropology,'" *Anthropology in Action* 12, no. 3 (2005), 12.

65. Cited in Price, *Anthropological Intelligence*, 225.

66. Cited in Ian Buruma, foreword to *Ruth Benedict, The Chrysanthemum and the Sword: Patterns of Japanese Culture* (New York: Houghton Mifflin, [1946] 2005), viii.

67. Geoffrey Gorer, *Japanese Character Structure and Propaganda* (New Haven, Conn.: Institute for Intercultural Studies, 1942), 12–13.

68. Weston LaBarre, "Some Observations on Character Structure in the Orient: The Japanese," *Psychiatry* 8 (1945): 326.

69. LaBarre, 326–27.

70. Price, *Anthropological Intelligence*, 39.

71. On the involvement with ethnographers in the Japanese Internment during World War II, see Orin Starn, "Engineering Internment: Anthropologists and the War Relocation Authority," *American Anthropologist* 13, no. 4 (November 1986): 700–720. For an early denunciation of American racism against the Japanese during World War II, see Gene Weltfish, "American Racism: Japan's Secret Weapon," in *Far Eastern Survey*, August 29, 1945, 233–37.

72. On Japanese "relocation," see Edward H. Spicer, "Anthropologists and the War Relocation Authority," in Goldschmidt, *Uses of Anthropology*, 217–38, 225.

73. Mead, *And Keep Your Powder Dry*, 3, 161.

74. Quoted in Deitchman, *Best Laid Schemes*, 33.

75. Quoted in Deitchman, 15.

76. Quoted in Deitchman, 181.

77. William Thomson, 1st Baron Kelvin, cited by David Burns, 3M Research

Lab, http://www.canada.com/calgaryherald/story.html?id=a6df4358-cec0-4555 -9efa-d7e66b4a31bc.

78. See the ESOC: Empirical Studies of Conflict site at https://esoc.prince ton.edu.

79. Deitchman, *Best Laid Schemes*, 246.

80. Cited in David H. Ucko, *The New Counterinsurgency Era: Transforming the U. S. Military for Modern Wars* (Washington, D.C.: Georgetown University Press, 2009), 53.

81. Robert S. McNamara, *In Retrospect* (New York: Random House, 1995), 32.

82. John L. Sorenson and David K. Pack, *Applied Analysis of Unconventional Warfare* (China Lake, Calif.: U.S. Naval Ordnance Test Station, 1964).

83. Deitchman, *Best Laid Schemes*, 246.

84. John C. Donnel and Gerald C. Hickey, "The Vietnamese 'Strategic Hamlets,'" *US Air Force Research Memorandum* (Santa Monica, Calif.: RAND, 1962), 7.

85. Donnel and Hickey, 27.

86. Sorenson and Pack, *Applied Analysis of Unconventional Warfare*, 13.

87. For the U.S. Department of Defense Project Maven, see https://www .govexec.com/media/gbc/docs/pdfs_edit/establishment_of_the_awcft_project _maven.pdf.

88. University of Maryland, "STOP Terrorism Software," Phys.Org, February 25, 2008, https://phys.org/news/2008-02-terrorism-software.html.

89. *Webster Dictionary*, s.v. "stochastic," https://www.merriamwebster.com/ dictionary/stochastic.

90. Aaron Mannes, et al., "Stochastic Opponent Modeling Agents: A Case Study with Hamas," in *Proceedings of the Second International Conference on Computational Cultural Dynamics* (2008), https://www.cs.umd.edu/~asliva/papers/ SOMAHamas-icccd08.pdf.

91. Mannes et al., "Stochastic Opponent Modeling Agents," 50.

92. Patrick Duggan, "Man, Computer, and Special Warfare," *Small War Journal,* January 4, 2016, http://smallwarsjournal.com/author/patrick-duggan.

93. For a provocative take on "tele-presence" written in a philosophical register, see Tom Cohen, "Introduction," *Telemorphosis: Theory in the Era of Climate Change,* vol. 1, 13–42 (Ann Arbor, Mich.: Open Humanities Press, 2012).

94. Project Maven, n.p.

95. Mannes et al., "Stochastic Opponent Modeling Agents," 53.

96. National Academies of Sciences, Engineering, and Medicine, *2015–2016 Assessment of the Army Research Laboratory* (Washington, D.C.: National Academies Press, 2016).

97. Kurzweil, "A Living Programmable Biocomputing Device Based on

RNA," *Kurzweil Weekly Newsletter,* July 28, 2017, http://www.kurzweilai.net/a-living-programmable-biocomputing-device-based-on-rna.

98. On the history of the concept of system, its use for integrating disciplines and genres, its bearing on the physically real, its falling out of favor in the nineteenth century, and its return in the twenty-first, see Clifford Siskin, *System: The Shaping of Modern Knowledge* (Cambridge, Mass.: MIT Press, 2017).

99. David Deutsch, *The Beginning of Infinity: Explanations That Transform the World* (London: Penguin, 2012), 302.

100. David Deutsch, "What Is Computation? How Does Nature Compute?" *A Computable Universe* (2012): 551–65.

101. Joint Chiefs of Staff, *The Department of Defense Dictionary of Military and Associated Terms,* JP 1–02, November 8, 2010, https://fas.org/irp/doddir/dod/jp1_02.pdf.

102. For the instrumentalist charge against the technical and "managerial vocabulary" in HTS theory, see Celeste Ward Gventer, David Martin Jones, and M. L. R. Smith, "Minting New COIN: Critiquing Counter-Insurgency Theory," in *The New Counterinsurgency Era in Critical Perspective,* ed. Celeste Ward Gventer, David Martin Jones, and M. L. R. Smith, 9–31 (London: Palgrave Macmillan, 2014), 19.

103. Siskin, *System,* 231, 233.

104. Roger Trinquier, *Modern Warfare: A French View of Counterinsurgency* (London: Praeger, [1961] 1964), 4–5.

105. Martin C. Libicki et al., *Byting Back: Regaining Information Superiority Against 21st-Century Insurgents* (Santa Monica, Calif.: RAND, 2007).

106. Cited in Roberto J. González, *American Counterinsurgency: Human Science and the Human Terrain* (Chicago: Prickly Paradigm Press, 2009), 81.

107. Paula Holmes-Eber, *Culture in Conflict: Irregular Warfare, Culture Policy, and the Marine Corps* (Stanford, Calif.: Stanford University Press, 2014), 5.

108. Holmes-Eber, *Culture in Conflict,* 6.

109. Philip Athey, "The Marine Corps Has Increased Troop Diversity with More Minorities and Women," *Marine Corps Times,* December 20, 2019, https://www.marinecorpstimes.com/news/your-marine-corps/2019/12/20/the-marine-corps-has-increased-troop-diversity-with-more-minorities-and-women-but-some-critics-say-its-not-pushing-future-growth/.

110. Holmes-Eber, *Culture in Conflict,* 174.

111. *Human Terrain System,* May 14, 2010, https://web.archive.org/web/20120426053324/http://humanterrainsystem.army.mil/Default.aspx.

112. The contents of the Wikileaks documents on the Human Terrain System are found at http://zeroanthropology.net/2010/07/27/human-terrain-teams-in-wikileaks-afghan-war-diary-raw-data/. For more on the gradual unraveling of the

HTS, see Maximilian C. Forte, "The Human Terrain System and Anthropology: A Review of Ongoing Public Debates," *American Anthropologist* 113, no. 1 (March 2011): 149–53.

113. Deitchman, *Best Laid Schemes*, 4.

114. Deitchman, 30.

115. For more on the failure of U.S. COIN practices in Vietnam, see D. Michael Shafer, *Deadly Paradigms: The Failure of U.S. Counterinsurgency Policy* (Princeton, N.J.: Princeton University Press, 1988), chapter 9.

116. Major Kevin R. Golinghorst, *Mapping the Human Terrain in Afghanistan* (Fort Leavenworth, Kans.: School of Advanced Military Studies, 2010), ii.

117. John Stanton, *US Army Human Terrain System, 2008–2013: The Program from Hell* (San Bernadino, Calif.: Create Space Independent Publishing Platform, 2013). A more affirmative presentation of HTS is offered (in the same year) in Christopher J. Lamb et al., *Human Terrain Teams: An Organizational Innovation for Sociocultural Knowledge in Irregular Warfare* (Washington, D.C.: Institute of World Politics Press, 2013).

118. Stanton, *US Army Human Terrain System*, 14, 12.

119. Louisa Kamps, "Army Brat," *Elle*, April 2008, 309.

120. Montgomery McFate, "Burning Bridges or Burning Heretics? A Response to González," *Anthropology Today* 23, no. 3 (June 2007): 21. For an overview of the arch of the Human Terrain System program from the beginning to its (alleged) end, see Paul Joseph, *"Soft" Counterinsurgency: Human Terrain Teams and US Military Strategy in Iraq and Afghanistan* (New York: Palgrave Macmillan, 2014).

121. McFate, "Burning Bridges or Burning Heretics?," 21.

122. Golinghorst, *Mapping the Human Terrain*, 30.

123. Golinghorst, 44.

124. Holmes-Eber, *Culture in Conflict*, 125.

125. Holmes-Eber, *Culture in Conflict*, 136.

126. Holmes-Eber, *Culture in Conflict*, 137.

127. David Petraeus, *Counterinsurgency Field Manual* (Chicago: University of Chicago Press, 2007).

128. On census work as part of COIN, see David Galula, *Counterinsurgency Warfare: Theory and Practice* (Westport, Conn.: Praeger, 2006), viii, 66, 74, 82.

129. See Derek Gregory, "The Everywhere War," *Geographical Journal* 117, no. 3 (September 2011): 238–50.

130. Anthony H. Cordesman, *Quadrennial Defense Review* (Department of Defense, 2010), http://csis.org/publication/2010-quadrennial-defense-review i, vi, and 8. The QDR is compiled every four years to put forward a twenty-year projection of U.S. military planning.

131. Jacob Kipp, "The Human Terrain System: A CORDS for the Twenty-First Century," *Military Review,* September–October 2006, 8–15, 4.

132. For further on green data, see Steve T. Mensh, "Electronic Warfare: Living in a Multi-spectral World," *Inside TS,* November 27, 2017, https://www.textronsystems.com/our-company/news-events/articles/inside-ts/electronic-warfare-living-multi-spectral-world.

133. For more on the promises and failures of MAP-HT, see González, *American Counterinsurgency,* 85–90.

134. Gventer, "Introduction," in *New Counterinsurgency Era.* For a critical take on the new COIN, see also Fred Kaplan, *The Insurgents: David Petraeus and the Plot to Change the American Way of War* (New York: Simon and Schuster, 2013).

135. Matthew Ford et al., "COIN Is DEAD—Long Live COIN," *Parameters* (Autumn 2012): 32–43.

136. On the still-living HTS program, after its purported cancellation in 2014, see Tom Vanden Brook, "$725M Program Army 'Killed' Found Alive, Growing," *USA Today,* March 10, 2016, https://www.usatoday.com/story/news/nation/2016/03/09/army-misled-congress-and-public-program/81531280/. On the Global Cultural Network, see Vanessa M. Gezari, "The Quiet Demise of the Army's Plan to Understand Afghanistan and Iraq," *New York Times Magazine,* August 18, 2015, https://www.nytimes.com/2015/08/18/magazine/the-quiet-demise-of-the-armys-plan-to-understand-afghanistan-and-iraq.html.

137. The Stalin dictum is used in T. X. Hammes's "Future War: Why Quantity Will Trump Quality," *Diplomat,* November 20, 2014, http://thediplomat.com/2014/11/future-war-why-quantity-will-trump-quality/.

138. See General David Petraeus, "Learning Counterinsurgency: Observations from Soldiering in Iraq," *Military Review,* January–February 2006, 2.

139. Department of Defense, FM-3.24 *Insurgencies and Counterinsurgencies,* May 2014, 139, https://irp.fas.org/doddir/army/fm3-24.pdf.

140. On war's "identity drivers," see David Kilcullen, "Countering Global Insurgency," *Journal of Strategic Studies* 28 (2005): 596–617.

141. Tom Blackwell, "Mapping 'White' Afghans Aim to End Civilian Deaths," *National Post* (November 8, 2008), https://zeroanthropology.net/2008/11/24/canadas-own-human-terrain-system-white-situational-awareness-team-in-afghanistan/.

142. On the *tor gund* and *spin gund* factions, see Christine Noelle, *State and Tribe in Nineteenth-Century Afghanistan: The Reign of Amir Dost* (New York: Routledge, 2016).

143. Kipp, "Human Terrain System," 3.

144. Michael Fitzimmons, *Governance, Identity, and Counterinsurgency:*

Evidence from Ramadi and Tal Afar (Carlisle Barracks, Pa.: US Army War College, 2013), 132.

145. *Instructions for American Serviceman in Iraq during World War II* (Chicago: University of Chicago Press, 2007), 5.

146. *Instructions,* 5.

147. U.S. Marine Corps, *Small Wars Manual,* FMFRP 12–15 (Washington, D.C.: U.S. Government Printing Office, 1940).

148. On Roosevelt's "Quarantine Speech," see John McV. Haight Jr., "Roosevelt and the Aftermath of the Quarantine Speech," *Review of Politics* 24, no. 2 (1962): 233–59.

149. Quoted in Maj. Ben Connable, "Human Terrain System Is Dead, Long Live . . . What?" *Military Review,* January–February 2018, non-paginated, https://www.armyupress.army.mil/Journals/Military-Review/English-Edition-Archives/January-February-2018/Human-Terrain-System-Is-Dead-Long-Live-What/.

150. Cited in Jenna Lark Clawson, *Ethical Landscapes of the Human Terrain System* (MA thesis, North Dakota State University, 2014), 60.

151. See references to Project Camelot's full name, Methods for Predicting and Influencing Social Change and Internal War Potential, in Joy Rohde, "Gray Matters: Social Scientists, Military Patronage, and Democracy in the Cold War," *Journal of American History* 96, no. 1 (June 2009): 199–224.

152. Department of the Army, *The US Army Human Dimension Concept,* TRADOC Pamphlet 525-3-7 (Fort Eustis, Va.: Department of the Army, 2014), 7.

153. Brian Michael Jenkins, *Stray Dogs and Virtual Armies: Radicalization and Recruitment to Jihadist Terrorism in the United States since 9/11* (Santa Monica, Calif.: RAND Occasional Papers, 2011), vii, https://www.rand.org/pubs/occasional_papers/OP343.html.

154. Stephen Graham, "US Military vs. Global South Cities," *Z Magazine,* July 20, 2005, www.zmag.org/content/showarticle.cfm?S.

155. Graham, 1.

156. Grégoire Chamayou, *A Theory of the Drone* (New York: New Press, 2015).

157. Chamayou, 16, 247.

158. See Andrew Cockburn, *Kill Chain: The Rise of High-Tech Assassins* (New York: Henry Holt, 2015), 32.

159. Graham, "US Military," 4.

3. WAR NEUROSCIENCE

1. Harold Hongju Koh, "The Obama Administration and International Law," U.S. Department of State, 2010, https://2009-2017.state.gov/s/l/releases/remarks/139119.htm.

2. Stephen Graham, "Geographies of Surveillant Simulation," in *Virtual Geographies: Bodies, Space, and Relations*, ed. Mike Crang, Phil Crang, and John May, 131–48 (London: Routledge, 1999), 133.

3. Koh, "Obama Administration," 5–6.

4. On the productive nature of analogy in science, see also Mary B. Hesse, *Models and Analogies in Science* (Notre Dame, Ind.: University of Notre Dame Press, 1966); Douglas R. Hofstadter and Emmanuel Sander, *Surfaces, and Essences: Analogy as the Fuel and Fire of Thinking* (New York: Basic Books, 2013); and Dedre Gentner, Keith James Holyoak, and Boicho N Kokinov, *The Analogical Mind: Perspectives from Cognitive Science* (Cambridge, Mass.: MIT Press, 2001).

5. See Derek Gregory, "Gabriel's Map: Cartography and Corpography in Modern War," https://geography.as.uky.edu/video/gabriel%E2%80%99s-map-cartography-and-corpography-modern-war. See also Eileen Rositzka, *Climatic Corpography: Re-mapping the War Film through the Body* (Berlin: De Gruyter, 2018).

6. Barbara M. Stafford, *Visual Analogy: Consciousness as the Art of Connecting* (Cambridge, Mass.: MIT Press, 1999), 9.

7. Keith J. Holyoak, Dedre Gentner, and Boicho N. Kokinov, "Introduction: The Place of Analogy in Cognition," in Gentner, Holyoak, and Kokinov, *Analogical Mind*, 2.

8. Douglas R. Hofstadter, "Epilogue: Analogy as the Core of Cognition," in Gentner, Holyoak, and Kokinov, *Analogical Mind*, 499–500.

9. Stephan Besser, "How Patterns Meet: Tracing the Isomorphic Imagination in Contemporary Neuroculture," *Confirmations: A Journal of Literature, Science, and Technology* 25, no. 4 (Fall 2017): 419.

10. Barbara M. Stafford, *Echo Objects: The Cognitive Work of Images* (Chicago: University of Chicago Press, 2007), 151.

11. On film and Deleuze, specifically the comparison of "rhizomatics" to probabilistic systems in brain processes, see Patricia Pisters, *The Neuro-Image: A Deleuzian Film-Philosophy of Digital Screen Culture* (Stanford, Calif.: Stanford University Press, 2012). Deleuze compares the brain and the screen more directly involving the question of speed in "The Brain Is a Screen: An Interview with Gilles Deleuze on the Time Image," *Discourse* 20, no. 3 (Fall 1998): 47–55. On the brain as a hardwired media transit system, see David Johnson Thornton, *Brain as Culture: Neuroscience and Popular Media* (New Brunswick, N.J.: Rutgers University Press, 2011). For Deleuze's theory of cinematic machines, see Gilles Deleuze, *Cinema 1: The Moving Image*, trans. Hugh Tomlinson and Barbara Habberjam (Minneapolis: University of Minnesota Press, 1986), 40.

12. Hava T. Sieglemann, "Neural and Super-Turing Computing," *Minds and Machines* 13 (2003): 103.

13. Sieglemann, 103.

14. Sieglemann, 104.

15. National Research Council, *Opportunities in Biotechnology for Future Army Applications* (Washington, D.C.: National Academy Press, 2001), 3.

16. Besser, "How Patterns Meet," 440. In addition to Bresser on the New Brain, see Davi Johnson Thornton, *Brain Culture: Neuroscience and Popular Media* (New Brunswick, N.J.: Rutgers University Press, 2011), 58–65.

17. Gabriel Robles-De-La-Torre, "Computational Modeling of Brain Function and the Human Haptic System at the Neural Spike Level: Learning Dynamics of a Simulated Body," in *The Oxford Book of Virtuality*, ed. Mark Grimshaw, 570–88 (Oxford: Oxford University Press, 2014), 570.

18. Robles-De-La-Torre, 581.

19. Robles-De-La-Torre, 582.

20. The ad hoc Committee on Military Intelligence Methodology for Emergent Neural Science is analyzed in James Giordano and Rachel Wurzman, "Neurotechnologies as Weapons in National Intelligence and Defense: An Overview," *Synthesis: A Journal of Science, Technology, Ethics, and Policy* (2011): 55–59.

21. Cited in Giordano and Wurzman, 56.

22. Cited in Giordano and Wurzman, 58.

23. On war research into biocybernetics in the 1970s, see R. A. Miranda et al., "DARPA Funded Efforts in the Development of Novel Brain-Computer Interface Technologies," *Journal of Neuroscience Methods* 244 (2015): 52–67.

24. National Research Council, *Opportunities in Biotechnology*, 4.

25. Ji-Wei Guo, "The Command of Biotechnology and Merciful Conquest in Military Opposition," *Journal of Special Operations Medicine* 9, no. 1 (Winter 2009): 70.

26. Committee on Military Intelligence, National Research Council, *Emerging Cognitive Neuroscience and Related Technologies* (Washington, D.C.: National Academies Press, 2008), 2.

27. Department of Defense, *Army Handbook: Commander's Guide to Biometrics in Afghanistan* 11, no. 25 (April 2011): ii.

28. Department of Defense, 5.

29. Department of Defense, 5.

30. Guo, "Command of Biotechnology," 72.

31. William James, "Bergson and His Critique of Intellectualism," in *Essays in Radical Empiricism; and a Pluralistic Universe* (New York: Dutton, 1971), 227.

32. James, 227.

33. Manuel DeLanda and Graham Harman, *The Rise of Realism* (Cambridge, U.K.: Polity Press, 2017), 11.

34. James, "Bergson," 252.

35. James, 228.

36. Henri Bergson, *Mind and Memory*, trans. N. M. Paul and W. D. Palmer (New York: Zone Books, 1991), 44.

37. Henri Bergson, *Mind Energy*, trans. H. Wildon Carr (London: Palgrave MacMillan, [1920] 2007), 187.

38. Bergson, 187–206.

39. James, "Bergson," 228.

40. Eugene Thacker, *Biomedia* (Minneapolis: University of Minnesota Press, 2004), 11.

41. James, "Bergson," 232–33.

42. James, 239.

43. Henri Bergson, *Creative Evolution*, trans. Arthur Mitchell (New York: Henry Holt, 1907), 323.

44. Gilles Deleuze, *Cinema 1: The Movement-Image*, trans. Hugh Tomlinson and Barbara Habberjam (London: Continuum, 2008), 1. There is a whole theoretical subgenre of writing on Delueze, film, and the brain, only some of which I use in the lead-up to posthuman war neuroscience. See, for example, Elizabeth Grosz, "Deleuze, Ruyer, and the Becoming-Brain: The Music of Life's Temporality," *Parrhesia* 15 (2012): 1–13; Andrew Murphie, "Deleuze, Guattari, and Neuroscience," in *Deleuze, Science and the Force of the Virtual*, ed. Peter Gaffney, 277–300 (Minneapolis: University of Minnesota Press, 2010); Lauren Berlant, "Intuitionists: History and the Affective Event," *American Literary History* 20, no. 4 (Winter 2009): 845–60; and William E. Connolly, *Neuropolitics: Thinking, Culture, Speed* (Minneapolis: University of Minnesota Press, 2002).

45. William D. Casebeer, *Natural Ethical Facts: Evolution, Connectionism, and Moral Cognition* (Cambridge, Mass.: MIT Press, 2003), 105.

46. Casebeer, *Natural Ethical Facts*, 105.

47. Bergson, *Mind Energy*, 193.

48. Kim Lachance Shandrow, "The US Military Wants to Inject People's Brains with Painkilling Nanobots," *Entrepreneur*, October 21, 2014, https://www.entrepreneur.com/article/238677.

49. Shandrow, n.p.

50. Shandrow, n.p.

51. Robert C. Jones, "Researcher Finds Better Way to Tap into the Brain," *News @theU*, March 19, 2021, https://news.miami.edu/stories/2021/03/researcher-finds-a-better-way-to-tap-into-the-brain.html.

52. Jones, n.p.

53. Kristen Houser, "DARPA Is Using Gamers Brain Waves to Train Robot

Swarms," The_Byte, February 8, 2020, https://futurism.com/the-byte/darpa
-gamers-brain-waves-train-robots-swarms.

54. V. I. Lenin, *Materialism and Empirio-Criticism: Critical Comments on a
Reactionary Philosophy* (Moscow: Foreign Languages Publishing House, [1908]
1952), 44.

55. On the idea of the enabling limit, see the definition of "protocol" presented by the Re: Enlightenment groups Oxford Protocols page, https://reen
lightening.org/protocols/.

56. Mark Solms and Oliver Turnbull, *The Brain and the Inner World: An Introduction to the Neuroscience of Subjective Experience* (New York: Other Press, 2002),
19.

57. I am aware of the ways in which realism and materialism do not overlap,
but to provide details of that discussion would be beyond the scope of this work.
For a taste of it, see chapter 1, "Realism and Materialism," in Manuel DeLanda and
Graham Harman, *The Rise of Realism*, 1–26.

58. Lenin, *Materialism and Empirio-Criticism*, 32, 38–40, 43, 46, 52, 58.

59. Norbert Wiener, *God and Golem, Inc.: A Comment on Certain Points Where
Cybernetics Impinges on Religion* (Cambridge, Mass.: MIT Press, 1966), 28.

60. Wiener, 29.

61. David Deutsch, *The Fabric of Reality* (New York: Penguin, 1998), 168.

62. Bergson, *Mind Energy,* 165.

63. Edward S. Casey, *Spirit and Soul: Essays in Philosophical Psychology,* 2nd
ed. (Putnam, Conn.: Spring Publications, 2004), 102.

64. Gilles Deleuze, *Cinema 2: The Time Image*, trans. Hugh Tomlinson and
Robert Galeta (Minneapolis: University of Minnesota Press, 1989), 210.

65. Deleuze, 331. For more on the relationship between topography and the
brain, see Elizabeth A. Wilson, *Neural Geographies: Feminism and the Microstructure of Cognition* (New York: Routledge, 1998). On cognition and corporeality
more generally, see Wilson's later book, *Psychosomatic: Feminism and the Neurological Body* (Durham, N.C.: Duke University Press, 2004); and Victoria Pitts-Taylor,
The Brain's Body: Neuroscience and Corporeal Politics (Durham, N.C.: Duke University Press, 2016).

66. Levi R. Bryant, *Onto-Cartography: An Ontology of Machines and Media*
(Edinburgh: Edinburgh University Press, 2014).

67. Tim Urban, "Neuralink and the Brain's Magical Future," Wait But Why,
April 20, 2017, https://waitbutwhy.com/2017/04/neuralink.html.

68. Deleuze, *Cinema 2,* 33.

69. Antonio R. Damasio, "How the Brain Creates the Mind," *Scientific American,* December 1999, 115.

70. Antonio R. Damasio, *Descartes' Error: Emotion, Reason, and the Human Brain* (New York: Penguin, 1994), 94.

71. Deutsch, *Fabric of Reality*, 99.

72. Damasio, *Descartes' Error*, 102.

73. X. Liu et al., "Quantifying Touch-Feel Perception: Tribiological Aspects," *Measurement Science and Technology* 19 (2009), https://www.researchgate.net/publication/228373791_Quantifying_touch-feel_perception_Tribological_aspects/.

74. Liu et al., n.p.

75. It should be noted that the kinds of brain images we see in scientific journals and popular magazines are not only composites of multiple neurological events edited into one picture but also often comprise data gathered from more than one brain. See Mara Cercignani, "Brain Microstructure by Multimodal MRI: Is the Whole Greater Than the Sum of Its Parts?" *NeuroImage*, November 4, 2017, https://www.sciencedirect.com/science/article/pii/S1053811917308868.

76. About the distinction between recollection and reality for Bergson, Casey reminds us that "when memory is taken in its recollective, fully imagistic mode ... the deremption between memory and imagination may not be so drastic as we first suspected." This, like the strict division between mind and matter, is another example of Bergson's "definitive overturn[ing] of dualism[s]." Casey, *Spirit and Soul*, 104–5.

77. Mark Humphries, "Mindlessly Mapping the Brain: Seeking Clarity for the Connectome," Spike, November 21, 2017, https://medium.com/the-spike/mindlessly-mapping-the-brain-1dec092a404d.

78. On 3D display screens, see Jason Geng, "Three-Dimensional Display Technologies," *Advanced Opt Photonics* 5, no. 4 (2013): 456–535, https://www.ncbi.nlm.nih.gov/pmc/articles/PMC4269274/.

79. Damasio, "Brain Creates the Mind," 114.

80. Damasio, 114.

81. See Anonymous, "Discovery of Quantum Vibrations in 'Microtubles' inside Brain Neurons Corroborates Controversial 20-Year-Old Theory of Consciousness," *Elsevier*, January 2014, https://www.journals.elsevier.com/physics-of-life-reviews/news/discovery-of-quantum-vibrations. This document is an abstract of the breakthrough Hameroff and Penrose article. See also Stuart Hameroff and Roger Penrose, "Consciousness in the Universe: A Review of the 'Orch OR' Theory," *Physics of Life Review* 11, no. 1 (March 2014): 39–78.

82. Fancis Crick and Christof Koch, "Towards a Neurobiological Theory of Consciousness," *Seminars in Neuroscience* 2 (1990): 263.

83. Crick and Koch, 263.

84. Colin Blakemore, "Rethinking the Senses," in *Science in Culture* (London: Arts and Humanities Research Council, 2014), https://ahrc.ukri.org/documents/case-studies/rethinking-the-senses-uniting-the-philosophy-and-neuroscience-of-perception/.

85. U.S. Army, TRADOC, "Mad Scientist Initiative," *Stand-To,* February 2018, https://www.army.mil/standto/archive/2018/01/30/.

86. U.S. Army, n.p.

87. U.S. Army, n.p.

88. Hameroff and Penrose, "Consciousness in the Universe," 41.

89. Jane Bennett, *Vibrant Matter: A Political Ecology of Things* (Durham, N.C.: Duke University Press, 2010).

90. Alfred North Whitehead, *Process and Reality: An Essay in Cosmology* (New York: Free Press, [1929] 1985), 108. Hereafter cited with page number in text as *PR.* According to Harman, Whitehead (along with Heidegger) pioneered a "new theory of objects in philosophy," and are therefore the primary informants of the speculative turn. See Graham Harman, *Towards Speculative Realism: Essays and Lectures* (Washington, D.C.: Zero Books, 2009), 22.

91. U.S. Army, "Mad Scientist Initiative," n.p.

92. Whitehead, *Process and Reality,* 19.

93. Whitehead, 28–29, 32, 340.

94. James, "Bergson," 249.

95. James, 231.

96. James, 249.

97. William James, "Proposing the Moral Equivalent of War," *Lapham's Quarterly,* 1906, https://www.laphamsquarterly.org/states-war/proposing-moral-equivalent-war.

98. James, n.p., emphasis in original.

99. James, n.p.

100. Carl von Clausewitz, *On War,* trans. Anatol Rapoport (New York: Penguin, 1982), 7.

101. Von Clausewitz, 7. For more on Clausewitz and war and means, see Mike Hill with Tom Cohen, "Black Swans and Pop-up Militias: War and the 'Re-rolling' of Imagination," in "War by Other Means," special issue, *Global South* 3, no. 1 (Bloomington: Indiana University Press, 2009): 1–17.

102. Manuel DeLanda and Graham Harman, *Rise of Realism,* 3.

103. Paul Virilio and Sylvère Lotringer, *Pure War* (New York: Semiotext(e), 1983), 30.

104. For a general overview of Virilio, see John Armitage, *Virilio and the Media* (New York: Polity Press, 2012); John Armitage, ed., *Paul Virilio: From Modernism to Hypermodernism and Beyond* (London: Sage, 2000); and, more to my purposes

regarding media and war, Nick Mansfield, *Theorizing War: From Hobbes to Badiou* (London: Palgrave Macmillan, 2008). For a more judicious review, see Douglas Kellner, "Virilio, War, and Technology: Some Critical Reflections," https://pages.gseis.ucla.edu/faculty/kellner/Illumina%20Folder/kell29.htm. About Kellner's critical reflections, more below.

105. Paul Virilio, *The Information Bomb* (London: Verso, 2000), title page.

106. Paul Virilio, *The Lost Dimension* (New York: Semiotext(e), 1991), 36.

107. Virilio, 36.

108. Paul Virilio, *Speed and Politics: An Essay on Dromology* (New York: Semiotext[e], [1977] 1986), 134.

109. Paul Virilio, *The Vision Machine* (Bloomington: Indiana University Press, 1994), 84.

110. Xu Liu et al., "Optogenetic Stimulation of a Hippocampal Engram Activates Fear Memory Recall," *Nature* 484 (April 2012): 381.

111. Editorial, "Scientists Want to Use Brain Implants to Tune the Mind," *Kavli Foundation*, Winter 2017, http://www.kavlifoundation.org/science-spotlights/scientists-want-use-brain-implants-tune-mind#.XB0sJCMrLzs.

112. Rachael Lallensack, "Lasers Turn Mice into Lethal Hunters," *Science*, January 12, 2017, https://www.sciencemag.org/news/2017/01/lasers-turn-mice-lethal-hunters.

113. For more of this kind of treatment from Virilio on virtual reality, see his *Open Sky* (London: Verso, 1997).

114. Paul Virilio, *War and Cinema* (London: Verso, 1988), 26.

115. Henri Bergson, *The Meaning of War: Life and Matter in Conflict* (New York: Diderot Classics, [1914] 2017).

116. Henri Bergson, *Two Sources of Morality and Religion* (London: Macmillan, 1935), 289.

117. Bergson, 291.

118. Bergson, 44.

119. Michel Foucault, *Society Must Be Defended: Lectures at the College de France, 1975–76*, trans. David Macy (New York: Picador, 1977), 89.

120. On this reduction of Bergson's antiwar philosophy to spiritualism, see Michael R. Kelly and Brain T. Harding, "Bergson's Theory of War: A Study of *Libido Dominandi*," *Philosophy and Social Criticism* 45, no. 5 (2018): 1–19, 2.

121. Kelly and Harding, 2.

122. On Bergson and spiritualism, see Jean Gayon, "Bergson's Spiritualist Metaphysics and the Sciences," in *Continental Philosophy of Science*, ed. Gary Gutting, published online January 1, 2005, https://onlinelibrary.wiley.com/doi/book/10.1002/9780470755501.

123. Defense Science Board, *Report of the Defense Science Board Task Force on*

Understanding Human Dynamics (Washington, D.C.: Office of the Secretary of Defense, 2009), xvi.

124. Defense Science Board, xvi.

125. Defense Science Board, xvii.

126. James Balm, "The Subway of the Brain: Why White Matter Matters," Biomedical Central, June 22, 2014, http://www.iapsych.com/im/brainsubway.pdf.

127. Balm, 1.

128. R. Douglas Fields, "White Matter Matters," *Scientific American* 298, no. 3 (March 2008): 55.

129. Almut Schüz, "Neuroanatomy in a Computational Perspective," in *The Handbook of Brain Theory and Neural Networks* (Cambridge, Mass.: MIT Press, 1995), 622.

130. Editorial, "Projecting a Visual Image Directly into the Brain, Bypassing the Eyes," *Kurzweil*, July 14, 2017, http://www.kurzweilai.net/projecting-a-visual-image-directly-into-the-brain-bypassing-the-eyes.

131. Julian Sanchez, "Revolutionizing Prosthetics," *DARPA*, July 9, 2018, https://www.darpa.mil/program/revolutionizing-prosthetics.

132. Larry Squire et al., eds., *Fundamental Neuroscience*, 3rd ed. (Amsterdam: Elsevier, Academic Press, 2008), 48.

133. Defense Science Board, *Report*, 12.

134. Patrick Caughill, "Researchers Have Linked a Human Brain to the Internet for the First Time Ever: Welcome to the 'Brainternet,'" *Futurism*, September 14, 2017, https://futurism.com/researchers-have-linked-a-human-brain-to-the-internet-for-the-first-time-ever.

135. On white matter and neuroplasticity, see Kae-Jiun Chang, Stephanie A. Redmond, and Jonah R. Chan, "Remodeling Myelination: Implications for Mechanisms of Neuroplasticity," *Nature Neuroscience* 19, no. 2 (February 2016): 190–97.

136. Gary Marcus and Jeremy Freeman, "Preface," in *The Future of the Brain: Essays by the World's Leading Neuroscientists*, ed. Gary Marcus and Jeremy Freeman, xi–xiii (Princeton, N.J.: Princeton University Press, 2015), xiii.

137. Saad Jbabdi, "Measuring Macroscopic Brain Connections *In Vivo*," *Nature Neuroscience* 18, no. 11 (November 2015): 1546.

138. Jbabdi, 1550.

139. Jbadbi, 1551.

140. Robin A. Miranda, "DARPA-Funded Efforts in the Development of Novel Brain-Computer Interface Technologies," *Journal of Neuroscience Methods* 244 (2015): 61.

141. Miranda, 61.

142. NIH, *The Brain Initiative*, January 2022, https://braininitiative.nih.gov (unpaginated).

143. NIH, "NIH BRAIN Initiative Unveils Detailed Atlas of the Mammalian Primary Motor," October 6, 2021, https://braininitiative.nih.gov/brain-pro grams/cell-census-network-biccn (unpaginated).

144. Kate Wheeling, "The Brains of Men and Women Aren't Really That Different, Study Finds," *Science,* November 30, 2015, http://www.sciencemag.org/ news/2015/11/brains-men-and-women-aren-t-really-different-study-finds.

145. Steven Smith et al., "A Positive-Negative Mode of Population Covariation Links Brain Connectivity, Demographics and Behavior," *Nature Neuroscience* 18, no. 11 (November 2015): 1565.

146. On the neurodiversity movement, see John Elder Robison, "What Is Neurodiversity?" *Psychology Today,* October 7, 2013, https://www.psychologytoday .com/us/blog/my-life-aspergers/201310/what-is-neurodiversity.

147. Presidential Commission for the Study of Bioethical Issues, *Gray Matters: Topics at the Intersection of Neuroscience, Ethics and Society,* vol. 2 (Washington, D.C.: Presidential Commission for the Study of Bioethical Issues, 2015), 13.

148. National Research Council, *Opportunities in Neuroscience for Army Applications* (Washington, D.C.: National Research Council, 2009), 25.

149. National Research Council, 25.

150. Turhan Canli, "Neuroethics and National Security," *American Journal of Bioethics* 7, no. 5 (November 5, 2007): 9.

151. National Research Council, *Opportunities in Neuroscience,* vii. See also Jonathan D. Moreno, *Mind Wars: Brain Science and the Military in the Twenty-First Century* (New York: Bellevue Literary Press, 2006), 27.

152. See "Brain Wave Frequencies," *Neurohealth,* January 4, 2022, https:// nhahealth.com/brainwaves-the-language/.

153. W. M. Tolles, "Self-Assembled Materials," *MRS Bulletin,* October 2000, 2.

154. Tristan McClure-Begley, "Targeted Neuroplasticity Training (TNT)," *DARPA,* December 16, 2018, https://www.darpa.mil/program/targeted-neuro plasticity-training.

155. Sara Reardon, "Memory-Boosting Devices Tested in Humans: US Military Research Suggests That Electrodes Can Compensate for Damaged Tissue," *Nature,* November 3, 2015, https://www.nature.com/news/memory-boosting -devices-tested-in-humans-1.18712.

156. Moreno, *Mind Wars,* 26–27.

157. Howard Caygill, "Physiological Memory Systems," in *Memory: History, Theories, Debates,* ed. Susannah Radstone and Bill Schwarz, 227–34 (New York: Fordham University Press, 2010).

158. The Royal Society, *Brain Waves Module III: Neuroscience, Conflict, and Security* (London: Royal Society, 2011), 8.

159. National Research Council, *Opportunities in Neuroscience,* 15.

160. N. Katherine Hayles, *Unthought: The Power of the Cognitive Nonconscious* (Chicago: University of Chicago Press, 2017), 1.

161. Press release, "HRL Receives DARPA Award to STAMP Learning into the Brain," *HRL Laboratories,* May 12, 2016, http://www.hrl.com/news/2016/05/12/hrl-receives-darpa-award-to-stamp-learning-into-the-brain.

162. Patrick Tucker, "The Military Is Using Human Brain Waves to Teach Robots How to Shoot," *Defense One,* May 5, 2017, https://www.defenseone.com/technology/2017/05/military-using-human-brain-waves-teach-robots-how-shoot/137622/.

163. DARPA, "Program Title: Neurotechnology for Intelligence Analysts (NIA)," 2008, https://www.esd.whs.mil/Portals/54/Documents/FOID/Reading%20Room/Science_and_Technology/08-F-0799_Neurotechnology_for_Intelligence_Analysts_NIA_2008.pdf.

164. IARPA, "Integrated Cognitive-Neuroscience Architectures for Understanding Sensemaking (ICArUS)," December 23, 2018, https://www.iarpa.gov/research-programs/icarus.

165. Hayles, *Unthought,* 2.

166. National Research Council, *Opportunities in Neuroscience,* 24, 27.

167. Brian Wang, "Halo Sport Brain Stimulators Accelerate Strength and Skill," *Next Big Future,* August 1, 2016, https://www.nextbigfuture.com/2016/08/halo-sport-brain-stimulator-boosts-and.html.

168. Brian Wang, "DARPA High-Resolution Neural Interfaces for Controlling Drones and Cybertech, *Next Big Future,* October 3, 2018, https://www.nextbigfuture.com/2018/10/darpa-high-resolution-neural-interfaces-for-controlling-drones-and-cybertech.html.

169. Richard A. Anderson, "Cognitive Neural Prosthetics," *Annual Review of Psychology* 61 (2010): 1–28.

170. See Jordan Pearson, "Brain Controlled Flight Is a Thing Now," *Motherboard,* May 28, 2014, http://motherboard.vice.com/read/brain-controlled-flight-is-a-thing-now.

171. Jason Dearen, "Mind-Controlled Drones Race to the Future," *Associated Press,* April 28, 2016, https://www.eng.ufl.edu/newengineer/news/mind-controlled-drones-race-to-the-future/.

172. See Lin Edwards, "Cyborg Beetles to Be the US Military's Latest Weapon," *Phys.org News,* October 15, 2009, http://phys.org/news174812133.html.

173. Connolly, *Neuropolitics,* 28.

Index

Cromwell, Oliver, 101

C3 model. *See* Command, Control, and Communications

Cuba, invasion of, 79–80

cultural IQ, 33, 34, 46

cultural issues, 35, 39, 75, 82, 115, 116, 118

cultural studies, 96, 118, 174, 175, 180

culture, 4, 19, 20, 22, 25, 34, 92, 93, 96, 103, 119–20, 124, 127, 178; Afghani, 117; anthropology and, 84–85, 86; biology and, 4, 85, 87, 100, 102, 108; demographic attitude and, 180; human body and, 98; human relations and, 125; machines and, 87; mapping, 121; materiality and, 95; mathematics and, 94; notion of, 125, 139; oral, 89; primitive, 86; print, 75, 89; race and, 97; technology and, 85; terrain-ing of, 140; visualizing, 120, 121; war and, 98, 126; weaponization of, 25, 132, 153, 181, 186

culture wars, 33

cyber-enabled counterinsurgency (CE-COIN), 26, 106–7, 108–9, 111, 112, 120; knowledge experiments within, 118; war anthropology and, 115

cybernetics, 3, 7, 14, 25, 30, 149, 158, 162, 174; breakthrough of, 175

Cybernetics 2.0, 8, 14, 174; netcentric war and, 9–12

cyberspace, 21, 163

Damasio, Antonio R., 157, 158, 159, 164, 176; brain and, 153; dispositional convergence zones and, 154;

mind–media–matter relation and, 154; quantum of, 155

Daniell, Robert, xvi

DARPA. *See* Defense Advanced sResearch Project Agency

data, xxi, 10, 27, 102, 103, 110, 126; accumulated, 104; big, 2; census, 115; cognition as, 148; discovering, 9; empirical, 92; ethnographic, 79; goals, 8, 15; green, 26; high-fidelity, 180; imperfection of, 103; innovative, 9; network-centric, 20; as physical transmission, 92–95; processing, 30, 88, 146–47, 175, 181; weaponization of, 108

data analysis, 102, 181

data points, 107, 122

data rights declaration, 12

death: dance of, 67; direct war, 189n3; life and, 162, 163

de-civilianization, xvii, xix, 2, 23, 39, 45, 48, 56, 107, 124, 182; post-human war and, xx

Dedrick, Calvert, 195n29

Defense Advanced Research Project Agency (DARPA), 29, 138, 145, 148, 182, 183, 185; biology and, 184; brain and, 175; cyber-beetle of, 186; microcomputers of, 146; MNEPs and, 148; neuroscientific projects of, 176

Defense Modeling and Simulation office, 114

Defense Science Board, 175

De Landa, Manuel, 11, 140, 141, 161

de la Puente, Manuel, 195n28

Deleuze, Gilles, 136–37, 152, 209n11, 211n44; Bergson and, 153;

MIKE HILL is professor of English at the University at Albany, State University of New York. He is the author of *After Whiteness: Unmaking an American Majority*, coauthor (with Warren Montag) of *The Other Adam Smith*, editor of *Whiteness: A Critical Reader*, and coeditor (with Warren Montag) of *Masses, Classes, and the Public Sphere*.